35 Advances in Biochemical Engineering/ Biotechnology

Managing Editor: A. Fiechter

Biotechnology Methods

With Contributions by
C. Bedetti, M. J. Beker, A. Cantafora,
E. Heinzle, Ch. S. Ho, A. I. Rapoport,
J. F. Shanahan, M. D. Smith

With 71 Figures and 29 Tables

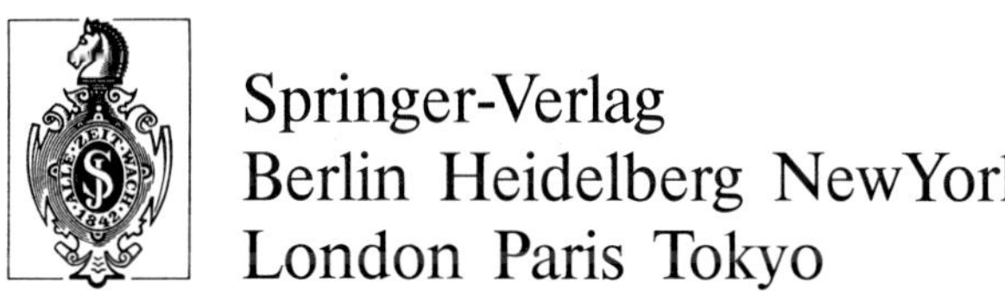

Springer-Verlag
Berlin Heidelberg NewYork
London Paris Tokyo

ISBN 3-540-17627-6 Springer-Verlag Berlin Heidelberg New York Tokyo
ISBN 0-387-17627-6 Springer-Verlag New York Heidelberg Berlin Tokyo

Library of Congress Catalog Coard Number 72-152360

Typesetting and Offsetprinting: Th. Müntzer, Bad Langensalza
Bookbinding: Lüderitz & Bauer, Berlin

2152/3020-543210

Table of Contents

Mass Spectrometry for On-line Monitoring of Biotechnological Processes
E. Heinzle 1

Extraction and Purification of Arachidonic Acid Metabolites from Cell Cultures
C. Bedetti, A. Cantafora 47

Carbon Dioxide Transfer in Biochemical Reactors
Ch. S. Ho, M. D. Smith, J. F. Shanahan 83

Conservation of Yeasts by Dehydration
M. J. Beker, A. I. Rapoport 127

Author Index Volumes 1–35 173

Mass Spectrometry for On-line Monitoring of Biotechnological Processes

E. Heinzle*
Institute for Environmental Research, Elisabethstr. 11, A-8010 Graz, Austria

1 Introduction ... 2
1.1 On-line Data Acquisition and Control: Measurement Technique Problems ... 2
1.2 Present State of Measurement Technique and Trends ... 3
2 Mass Spectrometry Applications in Fields Other than Biotechnology ... 5
2.1 Analysis of Gases and Volatiles ... 5
2.2 Isotopic Studies ... 5
2.3 Complex Mixtures ... 5
3 Principles of MS Operation ... 6
3.1 Vacuum ... 6
3.2 Ionization ... 7
3.3 Mass Separation ... 8
3.4 Detection ... 10
3.5 Data Handling ... 11
3.6 MS Control ... 13
4 Sampling Systems ... 14
4.1 Capillary Inlet ... 14
4.1.1 Design of a Capillary Inlet ... 14
4.1.2 Multiple Gas Streams ... 15
4.2 Membrane Inlets ... 17
4.2.1 Principles of Membrane Probes ... 17
4.2.2 Mass Transfer to and Through Membranes ... 17
4.2.3 Flow in Sampling Tubes ... 20
4.2.4 Design of Membrane Probes ... 21
4.3 Multiple Inlets ... 24
4.3.1 Transients of Switching ... 24
4.3.2 Steric Limitations ... 25
4.3.3 Control of Switching ... 25
5 Computer Aided Control of MS ... 25
6 MS Results of Bioprocess Monitoring ... 26
6.1 Gas Balancing ... 26
6.1.1 Estimation of Gas Reaction Rates ... 26
6.1.2 Error Analysis for Estimation of Gas Reaction Rates ... 29
6.1.3 Gas Balancing with Small Gas Streams ... 30
6.1.4 Examples ... 31
6.2 Dissolved Gases and Volatiles ... 31
6.2.1 Dissolved Gases ... 31
6.2.2 Monitoring of Volatiles ... 32
6.2.3 Fingerprinting ... 35
6.3 Further Methods ... 37
6.3.1 Increasing Volatility by Chemical Reaction ... 37
6.3.2 Pyrolysis ... 38
7 Future Trends ... 38
8 Concluding Remarks ... 39
9 Symbols and Abbreviations ... 39
10 References ... 41

* New adress: Chemical Engineering Laboratory, ETH-Zentrum, CH-8092 Zürich, Switzerland.

Advances in Biochemical Engineering/
Biotechnology, Vol. 35
Managing Editor: A. Fiechter

1 Introduction

1.1 On-line Data Acquisition and Control: Measurement Technique Problems

Biotechnological process investigation and operation is usually very expensive. Most of the processes are rather slow, often lasting a couple of days and aseptic operation is absolutely necessary to avoid competition from undesirable organisms. The media used may be very expensive, especially so in the case of animal-cell cultures which often require large amounts of fetal calf serum. For these and other reasons it is very desirable to get as much information as possible out of each experiment and to closely monitor or even control production processes in order to minimize costs and product quality. Product quality may be improved by increased selectivity, increased product concentration and increased reproducibility which may significantly reduce costs for down-stream processing.

Information may be collected by batch-sampling and subsequent chemical and/or physical analysis. This involves a number of manual operations but is most flexible and therefore the primary step to process analysis. For continuous collection of information this method is obviously not suitable and its applicability to control is limited to very slow processes and crude control methods. Automatic sampling can especially simplify over-night operation, but real improvement in process monitoring and control may be obtained by the introduction of automatic analysis. Automatic sampling increases risks of infection. Separation of liquid from cellular material and other solids still creates a lot of problems though much progress has been made recently [33,66,74,103]. For process control, on an industrial scale especially, the system described by Lenz et al. [66] seems to solve most of the problems. It involves filtration analysis for biomass estimation and on-line use of filtered samples for direct high performance liquid chromatography (HPLC) analysis.

Obviously the most desirable method for process monitoring is the application of continuously operating sensors. These should have a number of characteristics:

- The specificity should be high enough to give a useful signal for the concentration of one single chemical species or a group of them.
- Because of low concentrations, especially at the initial state of projects, the sensitivity must often be very high.
- In most cases steam sterilizability will be a necessary precondition.
- The sensor should not interfere with the process.
- Accuracy requirements will often be moderate (1–10% relative).
- Long term stability will reduce the need for recalibration, which under sterile conditions would in any case be very cumbersome.
- Sensor costs and maintenance requirements may be practical limits for certain applications.
- Dynamic characteristics will play an important role if the measured variable changes rather quickly. As a rule of thumb a sensor whose time constant is about 4–5 times larger than the time constant of the process itself can be used without significant problems. If the sensor has a first order delay characteristic the dynamic relative error would be $<2\%$.

To control a process the sensor will be an integral part of the feed-back control loop.

The quality of the sensor must be related to the process characteristics and the requirements for control performance. If the process is a completely black box, i.e. if no descriptive model exists, the variable to be controlled to desirable values must be measured directly. If some characteristics of the process are known, a model can be built which eventually allows the measurement and control of a related variable to finally control the desired variable. Estimators have been applied to allow calculation of interesting process variables [78,112,127,157]. Unfortunately in biotechnical processes the knowledge of the process mechanism is often very limited.

On-line estimation of variables may often be improved considerably by applying filter techniques [29,30,42,100,114].

1.2 Present State of Measurement Technique and Trends

A number of automatic analytical methods are now available on the laboratory scale as can be seen from a few examples listed in Table 1.

The performance of most of these methods is strongly dependent on sampling quality. In general all methods [6,33,63,66,74,82,119] are suitable for liquid separation, but dialysis membranes may be very fragile and some other filter probes may be subject to membrane fouling during long term operation. One method uses fresh filters for each sample [6,66] in order to avoid fouling. This probe, however, requires a large sample volume and cannot be easily used on a laboratory scale. Detecting methods may greatly vary (HPLC [6,33,66], autoanalyzer [119], IR [1], enzyme thermistor [28], MS [91] and others).

Conventional automatic analyzers would often need too large amounts of samples, which could be somewhat reduced by moving from segmented flow to flow injection

Table 1. Examples of automatic analysis in biotechnological processes

Variable	Method	Ref.
Biomass	Filtration probe	82)
Biomass, dissolved	Filtration probe and	6)
substrates and metabolites	HPLC	66)
Dissolved substrates and metabolites	Filtration cell and HPLC	33)
Ammonia, phosphate, glucose	Dialysis with enzymatic and chemical detection	119)
Penicillin, sugars	Enzyme thermistor	28)
Glucose	Dialysis with enzymatic detection	74)
Volatiles in the gas phase	Headspace/GC	23)
2-Oxoglutarate	Esterification/MS	89)
Glucose, ethanol, glycerol	FTIR/ATR	1)
Ethanol	Gas phase IR	80)
Extracellular enzymes	Ultrafiltration cell	63)

Table 2. Sensor for bioprocesses

Variable	Method	Ref.
Biomass; NAD(P)H	Fluorescence probe	[5,142]
Ethanol	Enzyme probe	[121]
Glucose	Autoclavable enzyme electrode	[21]
Penicillin, glucose	Enzyme electrode	[34]
Citric acid production	Redox-electrode	[4,62]
Methanol	Silicon tubing	[141]
Volatiles and dissolved gases	Porous Teflon tubing	[27,43,139,140]
Dissolved CO_2	Electrode	[88]
Dissolved O_2	Electrode	[52]

analysis. This also usually leads to a speeding up of analysis. Recent progress in miniaturization of biosensors leads us to expect significant progress in the field of continuous monitoring in a sampling stream, and may not require performance under aseptic conditions. The prospects in this field include miniaturized multicomponent probes which are not sterilizable and therefore very inexpensive and disposable. If monitoring of volatiles is of interest, detection in the gas phase may be much easier [23,80], but liquid-gas phase dynamics may cause errors in this case.

There are already a number of sensors that are used in bioprocess monitoring. These, however, are mostly sensors for physical variables and some of them are not directly steam sterilizable (Table 2). There exists a fairly long list of other sensors which are not sterilizable or subject to cross interferences (e.g. biosensors, ion-specific electrodes). Well established methods such as measurement of temperature, pressure, and pH are not listed here.

It can easily be seen that only a few sensors can directly measure product or substrate concentration under sterile conditions. Biomass concentration or activity cannot usually be measured on-line.

Present trends towards the development of new methods or for improving existing ones are classified in Table 3.

Table 3. Trends in on-line measurement

- Improvement in sampling techniques to couple powerful analytical methods to processes (HPLC [66], enzymatic sensors, automatic chemical analysis).
- Application of chip production technology to develop new miniaturized sensors. These may be multicomponent sensors being not steam sterilizable but very cheap. They can be used in very small amounts of sample stream to continuously monitor a series of variables in the non sterile region of a process.
- Adaptation of existing analytical methods (fluorimetry, laser technology [113], mass spectrometry, infrared spectroscopy [1]).
- Application of balancing methods to calculate new variables from available sensor signals [97].
- Incorporation of suitable mathematical filter techniques may further improve calculated variable accuracy [29,30,42,101,114].

2 Mass Spectrometry Applications in Fields Other than Biotechnology

2.1 Analysis of Gases and Volatiles

Mass Spectrometry (MS) has a long history dating back to since the beginning of this century. According to Brunnee and Voshage [14] the first quantitative gas analysis of volatile hydrocarbons was carried out in 1940. In 1949 MS had already been successfully applied to breath analysis [54] and later in chemical process monitoring [19,86,106]. Breath analysis is still an important field of on-line MS application [81]. Speed of detection and the inherent ability of MS to simultaneously analyze several gaseous compounds are the main advantages in this case. The application of MS for analysis of steel processes to monitor the progress of oxidation of carbon is a very well established method [19,106,111]. This example shows that MS can be used in a harsh industrial environment. In chemical processes MS may be superior to gas chromatography and other methods when speed of analysis is crucial e.g. in the production of ethylene oxide, propylene oxide, acrylonitrile, vinyl acetate and vinylchloride [25,26,37,83,108,109]. Multala et al. [79] applied MS analysis to computer control of distillation. Brockman and Anderson [13] used permeable membrane MS probes to monitor volatile products of electrochemical reactions.

In medical areas other than breath analysis, a great deal of effort has been applied to blood gas monitoring using membrane probes [11,64,126,137]. Monitoring of blood gases and volatile anaesthetics may be very attractive during critical operations. More recently non-invasive transcutaneous measurement of oxygen and helium dissolved in arterial blood using a specially designed membrane probe was described [40]. Weaver [130] gave a review on the medical and other applications of MS for measurement of volatiles. Langer et al. [65] measured dissolved gases in plants.

MS also has been applied in the monitoring of atmospheric pollution [75,84]. The greatest current use of MS for process controls is submarine atmosphere control [138].

2.2 Isotopic Studies

MS is ideally suited for monitoring isotopes. Radioactive isotopes however can usually be detected more easily and more sensitively by measuring radiation. Often it is advantageous to measure stable isotopes e.g. to avoid radiation hazards or for those elements which lack radioactive species. The most useful isotopes are ^{2}H, ^{13}C, ^{15}N, ^{17}O and ^{18}O [90,93,110].

2.3 Complex Mixtures

MS has also been applied in the direct analysis of complex mixtures though it is inherently limited in this respect because of overlapping of peak fragments from individual components. Despite this limitation, MS was successfully used to continuously analyse natural gas samples [123]. A general method was developed to improve analysis of mixtures [105]. MS was coupled to pyrolysis ovens to analyze coal [120], to classify microorganisms [2,135], to analyze polymers and polymer mixtures [136] and to analyze other complex biological samples [76].

3 Principles of MS Operation

The principles of MS operation are discussed in a number of text books [14,61]. In this article only a few important aspects with relevance to on-line analysis in fermentation will be discussed briefly.

Figure 1 shows schematically the units necessary for MS use in process monitoring.

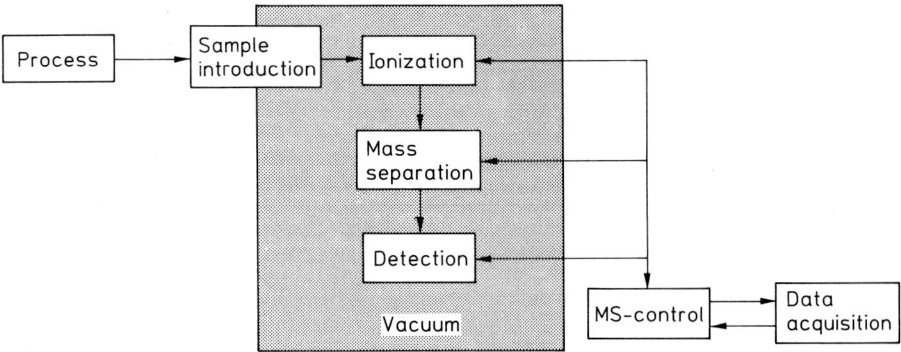

Fig. 1. Basic requirements for process MS

3.1 Vacuum

In MS, ions are used which will only survive sufficiently long if the vacuum is good enough. High pressure would allow collision with other molecules causing unpredictable reactions. The vacuum needed in the analyzer chamber will usually be $<10^{-5}$ mbar. Below this pressure the mean free path of particles will be sufficiently long and the number of collisions sufficiently small to allow good analysis (e.g. N_2: mean free path ≈ 0.5 m and the number of collisions of one molecule ≈ 70 per s, in this case the time from ionization to detection is < 10 ms).

The required vacuum may be produced by several methods: three will be briefly discussed. The most traditional and simplest is a combination of a rotary pump to produce about 10^{-3} mbar and an oil-diffusion pump to evacuate to the required working pressure. A turbomolecular pump may be used instead of the diffusion pump. Turbomolecular pumps give a very low contamination by hydrocarbons and other compounds. Oil-diffusion pumps need cold baffles to avoid diffusion of oil into the analyzer and this either necessitates cooling with liquid nitrogen, methanol dry-ice or the application of Peltier-cooling. Accumulation of condensed material can cause long term problems. Turbo pumps on the other hand, involve high rotation speeds of about 200 s^{-1}: their bearings will have a limited life-time, but prophylactic exchange of bearings after defined periods may avoid damage to the rotor and stator packages. Such damage would entail high costs. Start up to turbo-pumps is very fast and they can withstand rapid flooding with air. The pumping speed of the above pumps is mainly a function of the mass of molecules to be pumped. Sorption pumps do not rely on any movable parts except at start up. Their characteristics are also dependent on the chemical nature of the compounds to be pumped: noble gases in particular may be pumped at greatly reduced rates. The life-time of such pumps is limited by their sorption capacity and according to manifacturers will be about one year.

3.2 Ionization

A great number of ionization methods have been used [61] but for continuous operation and quantitative analysis electron impact ionization (EI) seems to be the most useful [108]. El can give stable operation over a long period of time [45]. One important requirement for obtaining stable operation is constant temperature [61]. As most biological samples may contain oxygen, proper selection of cathode material will enhance the life-time. A gas-tight ion source with cathode filament outside the ionization chamber but within the high vacuum region ($< 10^{-6}$ mbar) will increase the life-time of the filament. Because of this most process instruments have closed ion sources. With proper installation and selection of filament material, especially with respect to corrosion by oxygen, filaments can be continuously operated for one year or even more. Practically any chemical compound can be ionized by El provided the molecule can be introduced in the gas phase into the high vacuum. The energy applied to molecules is usually very high (70 eV) and therefore extensive fragmentation

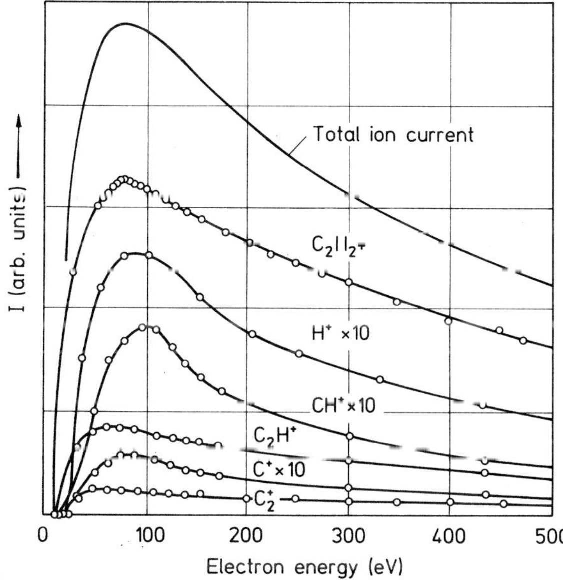

Fig. 2. EI ionization yield of acetylene as function of ionization energy [61]

Table 4. Ionization methods

Method	General applicability	Problems	Process use
Electron impact (EI)	Gas phase	Fragmentation, complex spectra	Extended and stable operation possible
Field ionization (FI)	Gas phase	Sensitive to dirt	Not suitable
Field desorption (FD)	Mainly solid samples	Only batch	Not suitable
Laser (LI)	Gas and solid phases		Generally possible
Photo (PI)	Gas phase		Generally possible
Thermal (TI)	Mainly solid samples		Not suitable
Thermospray (TS)	Liquids containing ions	Capillary sensitive to dirt and salts	Not suitable
Chemical (CI)	In the gas phase at higher pressure	Reactivity of the reagent gas depends on compound	Possible

is observed. Lowering the energy input favours higher masses which will be especially useful when analyzing mixtures. But fragmentation will still be considerable. In Fig. 2 ionization of acetylene is shown as a function of electron ionization energy.

It would be desirable to use soft ionization techniques which use much lower energies to ionize. These ideally would give only one single peak (e.g. molecular peak) from a single chemical species. A number of such soft ionization techniques are described in the literature and most of them also are commercially available. Recently it was claimed that chemical ionization has been developed to a state which allows process on-line monitoring [143]. Table 4 gives a list of ionization methods and a brief evaluation of their applicability for biotechnological process monitoring.

From Table 4 we can see that in the foreseeable future only El, photo-ionization and chemical ionization will be of practical importance in this field. Chemical ionization is very attractive, but obviously water greatly disturbs the reproducibility of measurement [109].

3.3 Mass Separation

At present two methods dominate on-line application, magnetic sector and quadrupole mass separation. Both methods are technically well developed and in most cases equally suitable.

Quadrupole Instruments

The quadrupole consists of four rods ideally of parabolic shape. A voltage consisting of a DC and a radio frequency component is applied. Opposite rods are connected. Positive ions entering the oscillating field will move on oscillatory paths. Only ions with a particular mass will move on stable paths, others with higher or lower masses will be discharged on collision with the rods of the quadrupole. Mass selection is achieved simply by setting the DC voltage e.g. between 0–10 V. Quadrupole instruments can be made very compact and sensitive. These instruments are ideally suited for very fast selected ion monitoring (SIM) or multiple ion detection (MID).

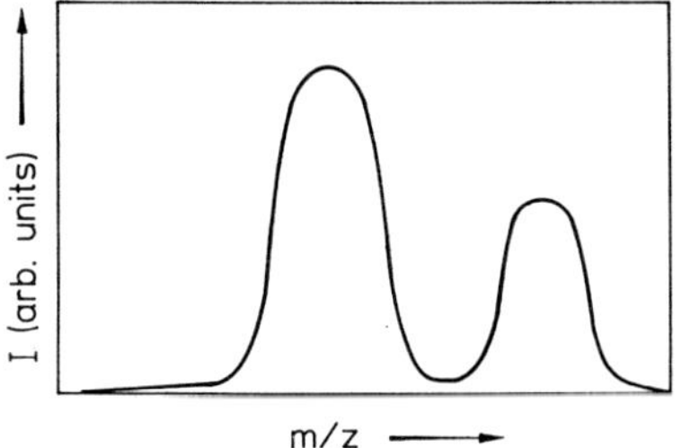

Fig. 3. Typical peak form of a quadrupole MS

In a quadrupole MS trapezoidal peaks cannot be obtained. The peak form can be manipulated to a certain degree by setting appropriate electric parameters. Figure 3 shows typical quadrupole peaks observed in practice. Using computer control of the MS slight disturbances in m/z (the mass to charge ratio) can be easily compensated for by occasional recalibration.

Quadrupole instruments are most frequently used in process gas analysis because they are cheaper and easier to use [109]. Quadrupoles of compact construction will allow operation at pressures up to 10^{-3} Torr [3] because a shorter mean free path of the molecules may be tolerated.

Magnetic Instruments

If a charged particle enters a magnetic field it will move on a circle whose radius is a function of the magnetic field, the particle velocity, and its mass to charge ratio (m/z). If the components to be measured are well defined and are not changing, fixed magnetic field and acceleration voltage combined with a number of simultaneously operating detectors may be used. More flexibility is obtained however, if scanning of either magnetic field strength or acceleration voltage is possible. Magnetic field scanning was improved by the introduction of Hall effect sensors. Both scanning methods are more complicated than changing the DC component of the quadrupole voltage. Magnetic instruments of appropriate construction can give higher resolution, and this resolution can be further improved by the addition of an electric field sector. Such a process instrument applied in trace gas analysis was recently introduced [138]. Generally magnetic instruments require more complicated ion sources because ions entering the magnetic field should have uniform energy. For quantitative analysis modern magnetic sector instruments are preferred [109] because the MS parameters can be selected to obtain trapezoidal peaks (Fig. 4).

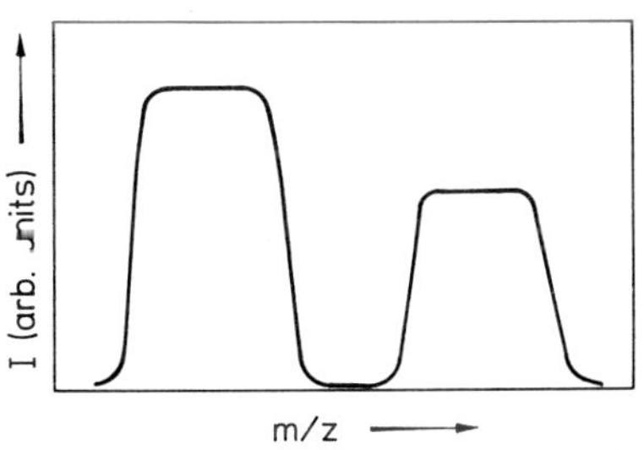

Fig. 4. Attainable trapezoidal peak form in magnetic sector instruments

This is possible if slits at the exit of the ion source and at the detector are properly selected. Because of the trapezoidal peak form, quantitative results are independent of small disturbances in m/z due to disturbances in acceleration voltage or magnetic field strength.

The greatest current use of MS for process control is for submarine atmosphere control [138]. These instruments were further developed recently by the addition of an electric sector to improve analysis of trace gases.

Other MS Types

Recently an ion trap detector especially designed for coupling to capillary-GC came on the market [36]. This very compact instrument may eventually be worth investigation for process applications because of its low susceptibility to changes in resolution by deposition of material on surfaces. Also, because of its compact construction, operation at slightly higher pressures is possible.

For future applications, Fourier transform (FT) MS may be very attractive. Smaller and cheaper instruments with their extremely good resolution may allow resolution of peaks which would usually overlap. Presently these instruments are too large and too expensive and sample introduction is critical.

Time of flight instruments seem to have certain advantages especially because of insusceptibillity to pollution but applications are very rare [25,26].

General Aspects

Resolution may play an important role in quantitative MS operation especially if high accuracy is required or if small peaks in the neighbourhood of large peaks have to be measured: the tail of the large peak may considerably influence the intensity of the smaller one. Resolution ($m/\Delta m$) e.g. at 10% valley is constant over the entire mass range in magnetic instruments. Quadrupole instruments are usually tuned to keep Δm constant. Δm will usually be 1, which means that neighbouring peaks of full masses are resolved over the whole mass range. This is obviously not possible if intensities differ very much. Quadrupoles usually give a certain mass discrimination at higher masses, which will be compensated during calibration.

Statistical errors of intensity measurement can be reduced by increased measurement time at the peak maximum. As the peak maximum is proportional to the peak area, (actually the basis for quantitative evaluations) integration of whole peaks is not necessary. But if measurement at the peak maximum cannot be guaranteed, whole peak integration can improve measurement accuracy in the absence of large neighbouring peaks.

3.4 Detection

Ions can be directly collected by a plate or Faraday cup to give an electric current which has to be amplified using an electrometer amplifier. Good long term stability, but relatively slow speed and limited sensitivity are the characteristics of such detection. Secondary electron multipliers (SEM) give additional very fast amplification of several orders of magnitude (10^4–10^8). This allows much faster operation and increased sensitivity of measurement. One major disadvantage of using an SEM is the aging process which especially intitially may cause a significant decrease in sensitivity. In our experience after some time of operation the SEM seems to become more and more stable. Ideally, automatic switching between SEM and Faraday cup will be used: this has been achieved in modern instruments [138].

SEM or channeltron multipliers can also be used for ion counting which is applied for high speed scanning with relatively small ion signals. The counting method has a very high dynamic range and is much less sensitive to noise originating from vibration or to drift because of aging of the detector. High intensities ($> 10^7$ ions s^{-1}), however, cannot be measured with currently available equipment [76].

3.5 Data Handling

Multiple Ion Detection

Ion signals collected from ion current measurements can be treated in a number of different ways depending on the purpose. The simplest method is recording of spectra or single peak intensities using an oscillograph or an analogue recorder. With microelectronics which usually will be used in modern instruments analogue signals will be digitized and can then be used in a number of ways. If selected ion monitoring is applied signals can be demultiplexed and a multichannel recorder can be connected to sample and hold circuits. Digital signals can also be used for computer treatment which may involve averaging, data storage and further mathematical treatment [20].

Analysis of Mixtures

If mixtures with superimposed ion fragment intensities have to be analyzed, a system of linear equations has to be solved to calculate concentrations of individual components in the mixture. The general method for quantitative analysis of mixtures with n components with concentrations $c_1, c_2, \ldots, c_n$ assumes linear superposition of peak intensities. The measured ion currents I_j for all masses are considered to be the sum of the contributions of all components, which are proportional to the concentration of each component, and to sensitivity coefficients $s_{j,1}, s_{j,2}, \ldots, s_{j,n}$:

$$\begin{aligned} I_1 &= s_{1,1}c_1 + s_{1,2}c_2 + \ldots + s_{1,n} \\ I_2 &= s_{2,1}c_1 + s_{2,2}c_2 + \ldots + s_{2,n} \\ &\cdot \\ &\cdot \\ &\cdot \\ I_m &= s_{m,1}c_1 + s_{m,2}c_2 + \ldots + s_{m,n} \end{aligned} \tag{1}$$

In matrix notation we can write

$$I = SC \tag{2}$$

In principle it is sufficient to measure $m - n$ suitable ion currents to determine C. The matrix S has to be determined by calibration with all the individual pure components. It can easily be seen that errors will be minimal if peaks are not overlapping, i.e. if only "pure masses" with a contribution from only one single component are measured.

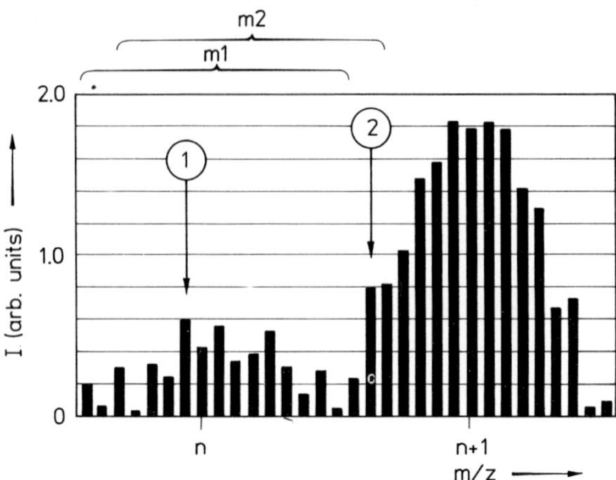

Fig. 5. Noisy measurement and detection of peak maxima. l: ion current; ml, m2: masks for maximum selection; 1,2: maxima found using masks ml and m2 respectively

If peaks are overlapping, several methods are described to reduce errors. Voogd et al. [123] first calibrated with pure components of natural gas getting absolute errors of 0–2%. After additional calibration with a suitable gas mixture similar to natural gas they calculated a correction factor for each component. Following this procedure the absolute errors were reduced to 0–0.2%. Schorr et al. [105] used statistical methods to calculate errors in analysis of gas mixtures and to minimize them. Breth et al. [12] showed that cracking patterns may be influenced by a number of parameters other than ionization method and energy.

Spectral Scanning

For scanning of whole spectra or parts of spectra, the maximum intensity for each mass unit has first to be selected from primary data already existing in digital form (e.g. 16, 32 or more measurements per unit mass). This can cause problems if instruments are not sufficiently well tuned or if the actual mass is in between two neighbouring integer masses. Especially in the lower intensity region and at high scanning speed, double peak detection can occur because of measurement noise (Fig. 5). If the mass is in between two neighbouring integer masses, the mask for maximum selection can be shifted. This can be especially critical if small peak maxima in the neighbourhood have to be found by automatic procedures. This is illustrated in Fig. 5, where a shift of the mask for maximum search gives different maximum selection with wrong m/z and intensity. In the case of measurement problems because of noise at low intensities, scan speed can be reduced or the threshold value for peak detection can be increased. This would lead to disappearance of smaller peaks thus reducing the information collected.

Sets of spectra can be analyzed using a number of methods. Library systems (which nowadays are also available for microcomputers) can be used to make attempts to identify single components or to compare spectra with those already stored in the library. These could be used as a data base for comparison of "fingerprints" of bioprocesses with each other.

Multivariate statistics may be used if pure components are not known and MS measurement data need interpretation for single components. A very powerful method in this respect is factor analysis [73,76,98].

Factor Analysis

Basically, factor analysis can be used for the following problem: A matrix of data **D** (e.g. mass spectra) has to be treated such that we find two matrices **C** and **R** which represent the concentrations and the spectra of the pure components:

$$\mathbf{D} = \mathbf{CR}\,. \tag{3}$$

D is a $c*r$ matrix with c different measurements and r masses; **R** is a $n*r$ matrix with the spectra of the n pure components; and **C** is a $c*n$ matrix with the concentrations of n pure components from c different measurements. **D** may be identical to the measurement ion intensities but usually a normalization would be necessary. The ion intensities may be normalized with the total ion current (I_{tot}) or with the maximum peak (I_{max}):

$$d_{i,j} = I_{i,j}/I_{tot} \quad \text{or} \quad d_{i,j} = I_{i,j}/I_{max}\,. \tag{4}$$

Heinzle et al. [49] used the water peak to normalize intensities in membrane probe "fingerprinting".

Using principal component analysis, a matrix **R** can be calculated which consists of a set of abstract eigenvectors each representing an abstract factor which may be identical to a mass spectrum of a pure component. Additionally, an associated set of abstract eigenvalues measuring the importance of each factor will be calculated. This corresponds to the concentration matrix **C**. A large eigenvalue indicates a major factor, whereas a very small eigenvalue indicates a less important factor.

Because of the introduction of errors in measurement and because of possible deviations from linear behaviour, the number of factors found in reality will always be c-1, where c represents the number of measurements. As the eigenvalues will be ordered with respect to their importance, an obvious discontinuity in this series of eigenvalues may indicate the number of factors actually determining the system. Malinowski [72] described an indicator function which can be used to determine the actual number of factors.

Factors extracted from principal component analysis will usually also contain negative peaks. For the identification of the components it may be useful to rotate the vectors to get spectra with better interpretability.

3.6 MS Control

The main purpose of MS control will be the control of the mass scan and the gain of ion amplification. For simple purposes, analogue circuits can be used to select certain masses or spectra to be recorded. Nowadays MS control can be performed using microcomputers which will give a very high flexibility for selecting certain masses or spectral regions [20].

Selection of masses is especially simple using quadrupole instruments. In this case usually a DC voltage of 0–10 V corresponding to the total mass range has to be supplied. With magnetic instruments either ion acceleration voltage, which will be in the kV region, or magnetic field have to be controlled. Compared to quadrupole instruments this is more difficult.

Microcomputers with suitable process interfaces can also be used to control inlet systems by switching valves [45].

4 Sampling Systems

A key role for the on-line application of MS in biotechnology is played by sampling systems. The problem is to reproducibly introduce samples from ambient pressure (usually about 1 bar) into the high vacuum of the MS ($< 10^{-5}$ mbar) This can be most easily done with gases. The introduction of other volatiles into the vacuum is usually more problematic. It is most difficult to continuously measure non volatile substances which actually form the major part of biologically interesting compounds.

4.1 Capillary Inlet

4.1.1 Design of a Capillary Inlet

For the analysis of gases, a capillary inlet will be the chosen method. As can be seen from Fig. 6, the pressure will be initially reduced from 1 bar to approximately 1 mbar using a capillary of an inner diameter of about 0.3 mm and a length of about 1 m. The necessary pressure drop is created by a rotary pump. Under these conditions flow in the capillary will be laminar. Flow in the connection line to the rotary pump also has to be in the laminar flow region. This avoids enrichment of compounds according to their masses, which only occurs in the molecular flow region where the mean free path of molecules is larger than the inner diameter of the tube.

Capillaries are usually made of stainless steel and can be heated electrically. At temperatures below the boiling point of the volatile compound condensation may occur if the vapor pressure is sufficiently low and (depending on temperature and the

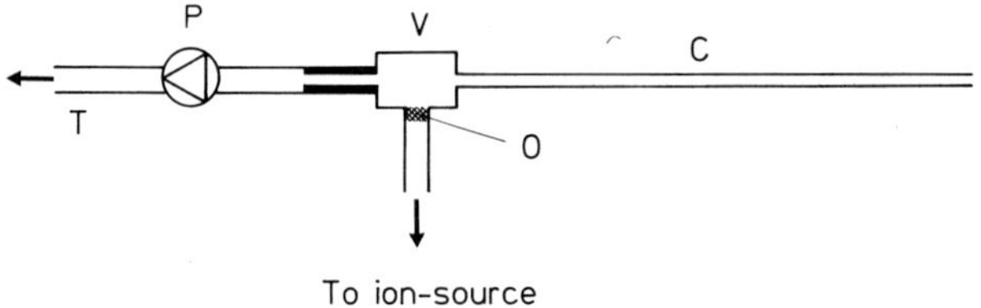

Fig. 6. Capillary inlet. C: capillary (i.d. ≈0.3 mm, length ≈1 m); V: valve; 0: orifice (10–20 μm) or sinter plate; T: vacuum tubing (i.d. >5 mm); P: rotary pump

Table 5. Response characteristics of capillary inlet (room temperature)[45]

Compound	$t_{90\%}$ [s]
Gases	<0.5
Ethanol	≈ 1
Acetone	< 1
n-Butanol	≈ 12
Water	≈ 4

material of construction) adsorption can lead to chromatographic effects and give dynamic errors. At very much higher temperatures catalytic effects causing chemical reactions of the compounds have to be expected. Capillaries also can be made of more inert materials like fused silica. This minimizes wall effects.

From the valve, where the pressure is about 1 mbar, the sample is expanded to the pressure of the ion source. The necessary flow resistance is supplied by either a narrow orifice of 10 20 μm or by a sintered steel or glass plate. Behind the orifice molecular flow occurs giving mass enrichment. If the ion source is gas tight then this enrichment is compensated for because molecules also leave this chamber under the molecular flow regime.

The dynamics of such capillaries is very fast for gases showing a mainly zero order delay of about half a second (Table 5) as determined by the time necessary for the flow through the capillary. At ambient temperature less volatile substances may show significantly slower reponses as can be seen from Table 5.

4.1.2 Multiple Gas Streams

From the speed of MS analysis it is evident that multiple gas streams may be analyzed using only one capillary inlet. It is only necessary to switch gas streams at the capillary entrance. Usually mixing delays in the sampling manifold will be the rate limiting step[15]. Though the capillary system responds very quickly to variation in gas partial pressures it can be seen from Table 5 that for less volatile substances the delay will be significant. If very high accuracy is required as may be the case for the partial pressure analysis of oxygen ($\leq 0.1\%$), dynamic errors can be significant and have to be under control. In this case the concentration difference between inlet and outlet of the reactor is very small.

The simplest way to manage the switching operation is to evaluate waiting times between measurement of different gas streams prior to actual analysis. Then switching and analysis can be made using a fixed frequency. Changes in the system dynamics influence the actual errors but will not directly be recognized.

A safer and more optimal method was used by Heinzle et al.[45]. It includes measurement of the dynamics of all masses of importance and introduction of a criterion for starting the actual measurement. A deviation d was calculated which by falling below a preset threshold value caused initialization of actual measurement.

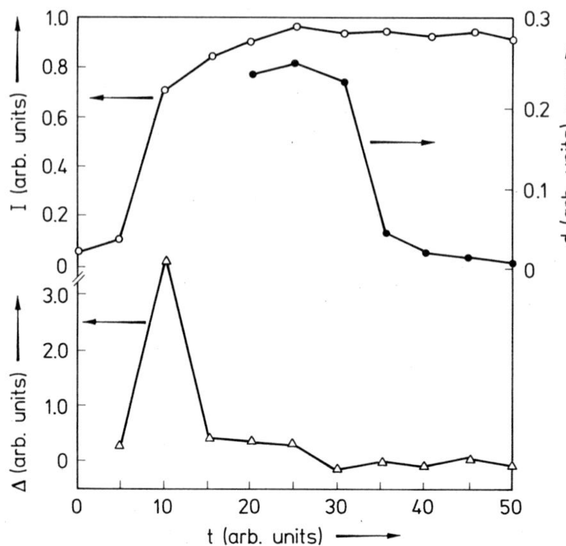

Fig. 7. Evaluation of the criterion for start of measurement. ○ measured ion current (I); $\Delta - (I_{n+1} - I_n)*(t_{n+1} - t_n)$; ● deviation d from Eq. (5)

$$d = \left| \frac{\sum_{n=1}^{m} (I_{n+1} - I_n)(t_{n+1} - t_n)}{t_m - t_1} \right| . \qquad (5)$$

The function of this criterion is illustrated in Fig. 7. The value of m was usually between 5 and 10. This simple criterion allows separation of the influences of noise and of signal trend. In certain cases it may be possible to normalize d to get d′.

$$d' = \frac{d}{1/m \sum_{n=1}^{m} (I)} \qquad (6)$$

Obviously this method is not suitable for rapidly changing concentration values. But in such systems where the concentrations change with approximately the same time constant as the dynamics of switching between sample streams, the method of multiplexing between sample streams is not applicable anyway.

For practical operation it was found that it is useful to define a maximum time limit for switching. In that case the computer should print out an error message and measure ion currents for diagnostic purposes. This or similar methods allow one to minimize measurement intervals and at the same time to maintain the dynamic error within preset limits.

MS producers nowadays offer complete systems including all software and hardware to measure gas partial pressures of a series of bioreactor gas streams. Fixed measurement intervals selected by the operator are usually used.

4.2 Membrane Inlets

In 1966 membrane inlets were introduced for blood gas monitoring [137] and also interfacing with gas chromatography (GC) of MS was perfected [99,128].

4.2.1 Principles of Membrane Probes

Membranes can give a sufficiently high flow resistance so as to directly separate samples at ambient pressure from the high vacuum of the MS: they also introduce selective transport mechanisms. Membrane permeability is mainly influenced by the following factors:
- temperature
- difference in partial pressure on the two sides of the membrane
- diffusivity of the chemical species in the polymer material
- solubility of the compound in the polymer material.

The last of these factors mainly reflects the chemical affinity between polymer and compound and is usually more important than diffusivity, which is mainly determined by the size of a molecule [55]. It can be expected that especially at high concentrations (or generally in non-ideal mixtures) fugacities are the important variables and not partial pressures. The situation, however, is even more complex because certain chemical compounds can have a considerable chemical affinity to the membrane and change its characteristics. Swelling may be observed. Molecules can also be expected to saturate binding sites of the polymer and therefore vary the solubility and consequently the permeability for other compounds. Furthermore, some of these processes may be much slower than others.

A great advantage of membranes can be their more or less pronounced selectivity for certain molecules. Polymers of interest usually do not allow penetration of ions or macromolecules. Most polymer materials are less permeable to water than to the organic molecules which would usually be present in culture fluids at much lower concentrations. Because of these characteristics, membrane probes always will need calibration for each analytical problem.

4.2.2 Mass Transfer to and Through Membranes

For practical purposes, porous and non-porous membrane types may be distinguished. An extensive survey of membrane transport is given by Hwang and Kammermeyer [55].

Non porous Membranes

The mass transfer through a membrane consists of the following steps:
1) diffusion through the stationary liquid boundary layer
2) sorption into the membrane
3) diffusion through the membrane
4) desorption out of the membrane
5) diffusion through the boundary layer

Fig. 8 schematically shows these steps and the important variables. In the case of a MS membrane probe, where the membrane separates process gas or liquid from the

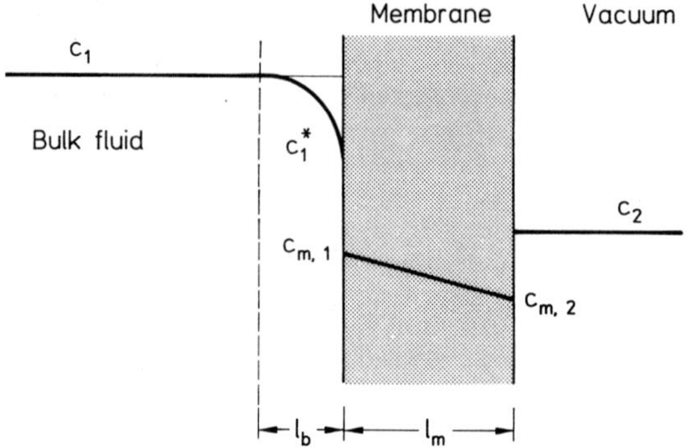

Fig. 8. Concentration profiles at the membrane interface. c_1: concentration in the bulk fluid; $c_1{}^*$: concentration at the liquid-membrane boundary; c_{m1}: concentration at the high pressure membrane boundary; c_{m2}: concentration at the vacuum membrane boundary; c_2: concentration in the vacuum; l_b: fluid boundary layer thickness; l_m: membrane thickness

high vacuum of the MS, steps 4 and 5 will not contribute significantly to the total flow resistance. Steps 1 and 2 cannot easily be analyzed separately, so usually these two resistances will be lumped together and under steady-state conditions we get for the flow rate F

$$F = a \frac{c_1 - c_1^*}{R_1} \tag{7}$$

where a is the membrane surface area, c_1 the concentration in the bulk, c_1^* the equilibrium concentration at the membrane surface, and R_1 the film resistance at the high pressure side. R_1 will be a function of the diffusion coefficient within the liquid and into the membrane, and of the boundary layer thickness. This thickness will be mainly determined by hydrodynamic conditions and by the viscosity of the fluid [102]. In the case when the fluid is liquid, the membrane separation process is usually called pervaporization.

For the diffusion within the membrane under steady-state conditions Fick's first law may be used to define the flow using concentrations at the membrane boundaries (c_{m1}, c_{m2})

$$F = Da \frac{c_{m,1} - c_{m,2}}{l_m} \tag{8}$$

D is the diffusion coefficient in the membrane and l_m the thickness of the membrane. Membrane probes should be designed such that the flow resistance within the membrane itself is the rate limiting step. This will avoid dependencies on hydrodynamic conditions on the process side of the membrane probe. Problems connected with dissolved oxygen measurement, where the liquid boundary layer plays an important role were discussed by Heinzle et al. [51].

Even with pure liquids the diffusion coefficient in the membrane may be a function of concentration (D(c). This is often described by an exponential function

$$D = D_0 e^{\alpha c} . \tag{9}$$

D_0, the diffusion coefficient at zero concentration, and α are both functions of temperature and the chemical and physical nature of the polymer and solvent.

The equilibrium contration in the membrane can often be described using Henry's law

$$c_1 = Hp_1^* . \tag{10}$$

In any membrane transport process temperature plays an important role. Permeability may be increased or decreased as can be seen from Fig. 9 which shows permeability of gases in several polymer materials. At the phase transition temperature (change in degree of crystallinity) breaks are observed. Permeation of real liquid mixtures is far from ideal permeation due to possible modification of the microstructure of the polymer film as well as due to mutual interaction among liquid permeants within the film [35,55,67,95].

Prime candidates for diffusive membranes suitable for MS membrane probes are organic polymer materials. Possibilities for the synthesis of membranes with

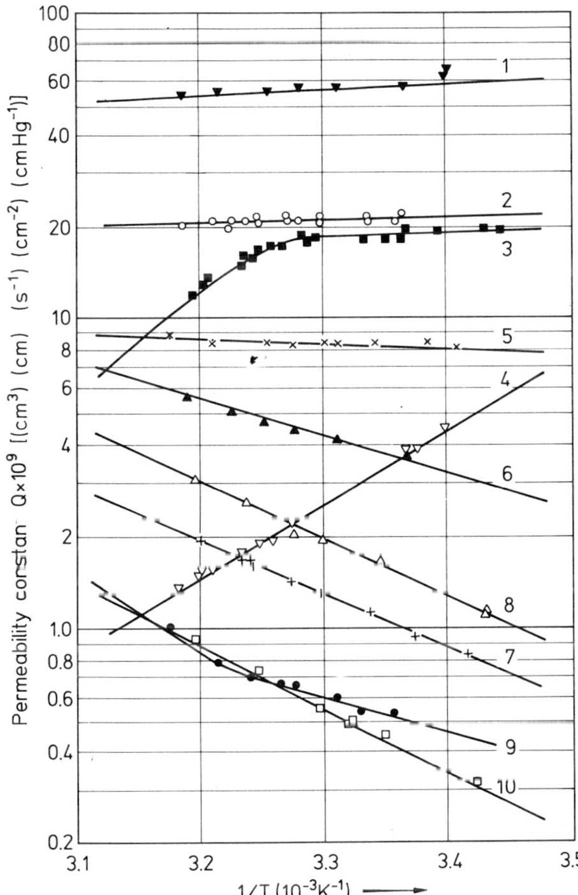

Fig. 9. Gas permeability of plastic films. 1: H_2, vinyl film, commercial calendered, no directional pattern; 2: CO_2, same as 1; 3: H_2, vinyl film, experimental press polished, no directional pattern; 4: CO_2, same as 3; 5: H_2, vinyl film, commercial calendered, no directional pattern; 6: CO_2, same as 5; 7: H_2, polyethylene film, commercial, extruded, directional pattern; 8: CO_2, same as 7; 9: H_2, vinyl film, experimental, extruded, no strain pattern; 10: CO_2, same as 9. (From Ref. [55])

desired chemical and physical properties are numerous. Often steam sterilizability is required which drastically reduces the number of available polymer materials. In most cases measurements will have to be made in an aqueous environment. Large amounts of water are not suitable for vaccuum applications. Membranes with a certain hydrophobicity will be required to allow enrichment of desired molecules. Membranes applied should not be susceptible to extensive swelling either by water or organic molecules in the measured fluid. Swelling will change transport selectivity usually resulting in a reduction of the separation factor. If the concentration of the material causing the swelling varies, slow variations in calibration factors have to be expected. Such variations would not allow reliable quantitative measurements to be made.

Porous Membranes

Porous membranes will have pores with diameters larger than the solvent or gas molecule sizes. Depending on pore size, temperature and pressure, flow is in the laminar, molecular or intermediate flow regime. At medium and especially at higher pressures adsorption effects are of increased importance. Porous membranes will usually have too little flow resistance to be suitable for one step separation between process and high vacuum of the MS. They may however be useful for special purposes as was suggested by Heinzle et al. [45] for porous Teflon membranes. With such membranes gas bubbles touching the membrane can be directly sucked through, whereas water can only penetrate as water vapour.

4.2.3 Flow in Sampling Tubes

Gaseous fluids can flow under three extreme regimes — laminar, turbulent and molecular. Turbulent flow of gases at reduced pressure is very unlikely and therefore not considered here. In the laminar flow region the mean free path of molecules in the gas phase is smaller than the dimensions of the vessel whereas in the molecular flow regime the mean free path is larger. In the laminar flow region, flow velocity is not a function of the mass of the molecules. Dispersion only occurs because of the parabolic velocity profile. Usually wall effects will play a minor role. Wall effects may however be significant in tubes which are very long in comparison to their diameter, and with sufficient chemical affinity between wall and components of the gas. Higher temperatures will increase the viscosity of gases, and at the same time may support catalytic reactions at the wall surface. Proper selection of inert materials (Teflon, fused silica) reduces adsorptive effects.

Under the molecular flow regime, small molecules will travel faster than big ones as can be seen from the well known formulation of Knudsen [55].

$$F = \frac{8\pi r_t^3}{3(2\pi RMT)^{1/2}} \frac{p_1 - p_2}{l_t} \tag{11}$$

The flow rate F is dependent on the radius of the tube r_t, on the pressure difference between inlet p_1 and outlet of the tube p_2, which is the driving force of flow, on temperature T, on the length of the tube l_t, and on the molecular weight M. R is the gas constant.

In tubes between membrane probes and the ion source, flow is expected to be molecular or in the transient region between molecular and laminar flow. This will cause an enrichment of molecules depending on their mass. This problem will certainly increase with the length of a tube. Under this flow regime molecules will very frequently hit the walls and thus have sufficient opportunity to adsorb; this is the first step in an eventual catalytic reaction. As adsorption is an exothermic process, increased temperature will force the thermodynamic adsorption equilibrium in the direction of desorption. Increased temperature will on the other hand also speed up possible catalytic reactions at the walls.

In coupling capillary GC to MS, the direct transfer line leading total eluent, including carrier gas, to the ion source will ideally be made of fused silica to give minimum disturbance at temperatures even in the region of 300 °C. Coating materials used are usually sensitive to oxygen at elevated temperatures. Evidently this consideration applies to the analysis of all aerobic fermentation samples.

For applications in industrial environments it is desirable to use materials that are robust. Heinzle et al.[45] simply used copper tubing (i.d. 4 mm). Problems were encountered when acetoin and butanediol were measured. Stainless steel tubes were connected with flexible portions to avoid direct transmission of vibrations to the MS, which would disturb measurements. For the measurement of these volatiles, thick walled Teflon tubing was found to be best amongst the three materials tested. Glass and fused silica tubing (i.d. 4 mm) is not flexible and seems only to be well suited for laboratory experiments where mechanical problems can be controlled much more easily.

As expected, an optimum temperature was found for each material where a maximum amount of unreacted molecules passes through the tube. Figure 10 shows results with copper tubing and measurement of acetoin dissolved in water. With stainless steel, Teflon, and glass the maximum is much flatter.

Bohátka et al.[7,8] used stainless steel tubing (length = 7 m; radius = 0.8 mm). With this tube they reported a response time $\tau_{63\%}$ of about 3 min at 90 °C.

4.2.4 Design of Membrane Probes

The basic design of a membrane probe is very simple as can be seen in Fig. 11.

The membrane (usually an organic polymer material) is sealed onto a tube which directly leads to the ion source of the MS. Usually the membrane which may be

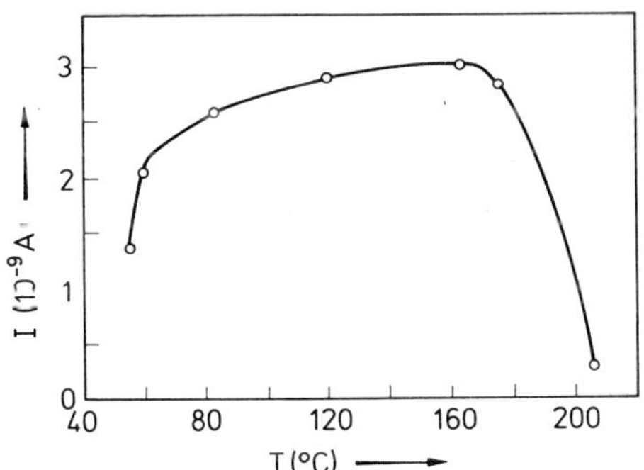

Fig. 10. Measurement of acetoin at various temperatures of the gold plated copper tube connection. 1: ion current of the molecular peak of acetoin (m/z = 88)

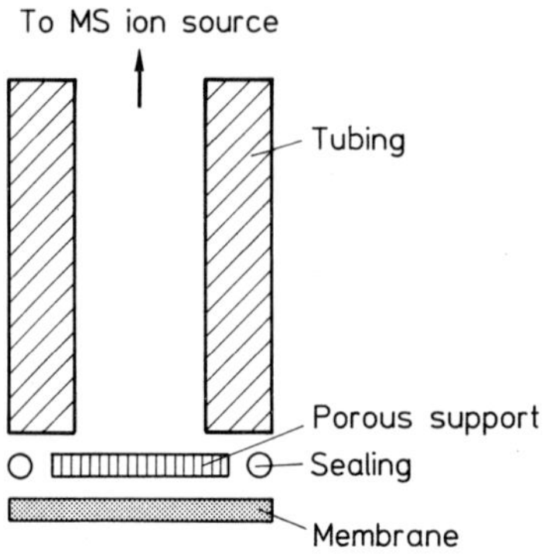

Fig. 11. Basic design of MS membrane probe

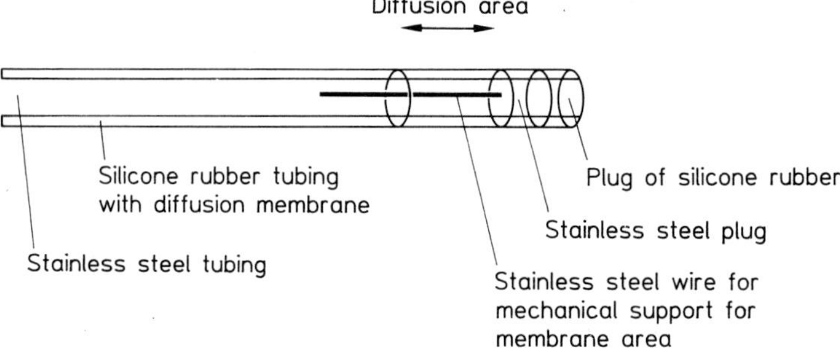

Fig. 12. Catheter membrane probe [11]

very thin and susceptible to mechanical stress needs a support. Following this basic design a series of modifications have been used and are described in the literature. In addition to the design of the probe itself a great deal of effort has been expended on the optimal design of hydrodynamic conditions at the membrane surface, because transport to the membrane surface may play an important role as discussed above.

Woldring et al. [137], Brantigan et al. [11] and subsequent authors describe membrane probes for the measurement of dissolved gas concentrations in blood vessels. In the design of Brantigan et al. [11] a silicone rubber membrane was supported by a thin (i.d. 0.5 mm) stainless steel capillary (Fig. 12). The outside diameter of the final sensor was 1 mm. The total diffusion area was about 43 mm^2. The response time of the probe ($\tau_{63\%}$) was about 30 s.

Johnson et al. [58] used a similar design but replaced the stainless tubing by a more flexible Sarane tubing (polyvinylidine chloride). Despite the obviously higher permeability of the polymer tubing compared to stainless steel, mass transfer through the tube walls was still a factor of about 100 times smaller than mass transfer through the silicone membrane. Obviously this increased background signal limits measurements in the low concentration region. The response characteristics for O_2, N_2, and CO_2 were similar to those of stainless steel tubing of the same length showing a time ($\tau_{63\%}$) of approximately 30 s.

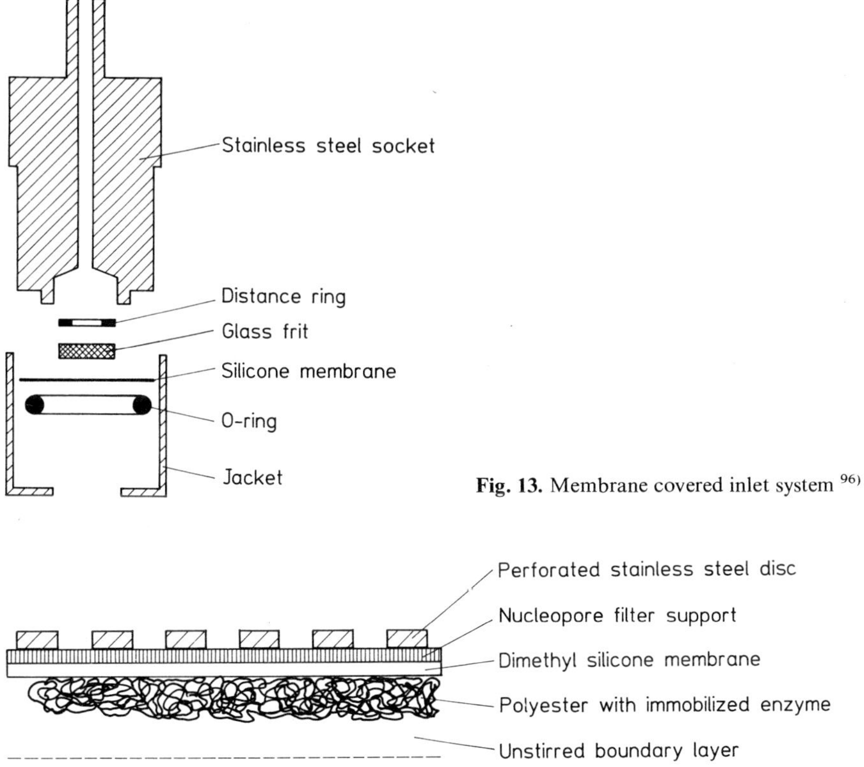

Fig. 13. Membrane covered inlet system [96)]

Fig. 14. Immobilized enzyme membrane probe [129)]

Reuss et al. [96)] who measured dissolved oxygen, carbon dioxide and methanol in the actual media used a somewhat modified design suitable for application in bioreactors (Fig. 13).

A similar design of membrane probe using 50–200 μm silicon rubber membrane and a glass fritt support was used by Pinnick et al. [85)] for monitoring volatile electrochemical products. Heinzle and Lafferty [48)] described a probe with only small diffusion area having no mechanical support. A slightly modified sensor with convex membrane surface to avoid possible dead zones was described by Heinzle et al. [45)]. Schmidt et al. [103)] applied silicone paste onto a sintered porous support material to get polymerization in the support. Westover et al. [134)] used silicone rubber hollow fibre probes to analyze traces of organic volatiles in water and in air. Weaver et al. [132)] combined a membrane probe with immobilized enzymes. The design of the enzyme-membrane system is shown in Fig. 14. With a more complicated sandwich structure of several membranes, problems in dynamic behaviour were found which were attributed to bulging of some of the membrane layers.

A number of authors [24, 60, 70, 71, 87, 89, 93, 131)] describe combinations of membrane probes of the design already described with measurement or reaction chambers, using various methods for the generation of reproducible turbulence.

Especially if very thin membranes are used it is strongly advisable to install safety shut off of the high vacuum system. Heinzle et al. [45] used fast total pressure measurement with an upper pressure limit of about 10^{-5} mbar. A pressure increase above this set point caused immediate shut-off of the filament and closing of all vacuum connections. Failures with thin membranes proved safe operation of this system. Spent media could only enter tubing lines, which could quite easily be cleaned. It was not necessary to turn off the vacuum of the MS.

4.3 Multiple Inlets

From the basic knowledge of operation of MS it can be deduced that not only single inlets but two or more can be connected to one MS. Besides connection of a capillary inlet membrane probes can be coupled to the ion source of the MS. Automatic valves, either magnetic or pneumatic, can be used to automatically switch between different inlets. There are, however, two main limitations to the number of inlets:

— frequency limits because of transient delays
— steric limitations

4.3.1 Transients of Switching

Each connection or disconnection of an inlet will cause a significant pressure variation. For measurement, steady state conditions have to be attained to allow useful results to be collected. The time necessary to reach steady conditions depends very much on pressure, on the kind of substance introduced, on the design of the vacuum system and on the temperature of the MS system. Higher pressure will usually favour faster transients because of the reduced wall influences.

Pressure, however, is limited because of the principles of vacuum physics. Miniaturization of ion source and mass separation would guarantee a sufficient mean free path of molecules at even higher pressure. Present limitations are usually of the order of $p_{tot} \leqq 10^{-5}$ mbar, but new designs of MS with smaller dimensions allow operation up to 10^{-3} mbar [3]. Another limitation for operation at higher pressure is the life-time of the filament. As described earlier, this problem can be solved by application of gas-tight ion sources or by ion-trap MS.

Transient speed is reduced by substances with an enhanced tendency to adsorb on the walls of the MS. Water and a number of organic molecules will create problems in this respect. This problem can be minimized by increasing the temperature, which accelerates adsorption and desorption processes. This allows equilibrium conditions to be attained much faster. There are, however, practical limits to temperature elevation. High-vacuum sealings on the market can tolerate 400 °C but automatic valves to operate at this temperature are very expensive. As has been shown above, high temperatures may reduce sensitivity.

Optimal design of the vacuum system can minimize switching delays. High vacuum regions should have open structure so as not to introduce high flow resistances. Membrane probes in the idle position should be kept at the same pressure as during the measurement. Heinzle et al. [45] connected the membrane probes to the centre of the rotor-stator package of the turbopump, where the

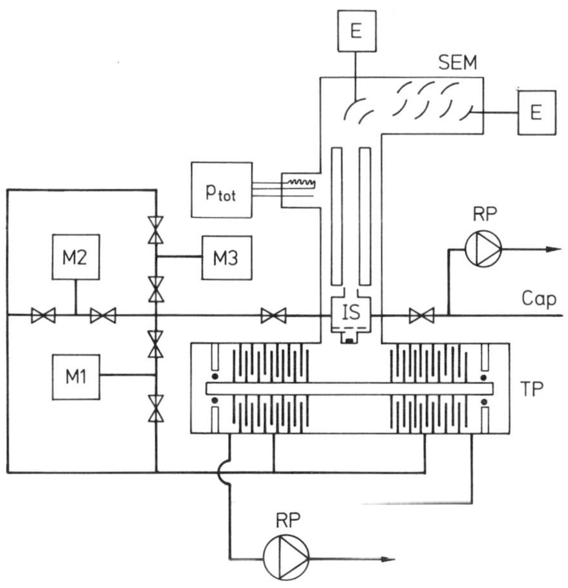

Fig. 15. Membrane probe connection to quadrupole MS [45]. M1–M3: membrane inlets: E: electrometer amplifier; SEM: secondary electron multiplier; p_{tot}: total pressure measurement; IS: ion source; TP: turbomolecular pump; RP: rotary pump; Cap: Capillary inlet

pressure was between 10^{-3} and 10^{-5} mbar (Fig. 15). Introduction of suitable apertures guaranteed uniform pressure at all inlets and this greatly reduced problems with switching gradients.

4.3.2 Steric Limitations

As high-vacuum tubes should be kept short to allow fast operation, it is not very likely that a whole series of membrane probes in a typical pilot plant will be connected to one MS. Bohátka et al. [8] and Szilágyi et al. [115] described the use of MS membrane probes with several meters tubing length and a couple of measurement points.

4.3.3 Control of Switching

Control of switching operation can be done using similar criteria as for switching between a number of gas streams. Either a fixed switching program can be applied or switching can be made on the basis of MS measurement results. The latter method will guarantee error limitations and at the same time minimize waiting periods between measurements.

5 Computer Aided Control of MS

As has been stated in previous sections, control of MS mass scans and amplification of ion currents is best be achieved using microprocessors. For fast scanning it would be desirable to have higher A/D conversion rates than 10 kHz. With microprocessors presently available it is ideal to dedicate one fast processor to mass scanning and ion current detection. This processor should have fast programs to find maxima of peaks and to reduce data to a two dimensional array containing mass numbers of peak maxima and corresponding intensities. Eventually peak integration may also be incorporated into these programs. Selected ion monitoring (SIM) with avaraging of ion intensities should also be available. MS manufacturers nowadays will offer such equipment including "user-friendly" interfacing. A second microcomputer should control this MS microprocessor demanding measurements and and collecting pretreated data. This computer should have extended disk storage capacity to store measured data ($\geqq 10$ Mbytes). In combination with the necessary computer process interfaces control of inlet stream selection should be possible. These interfaces should also allow additional measurements such as gas flows, dissolved oxygen concentration, pH etc. and the activities of control elements on processes. Additional hardware should include graphic output devices (plotter, graphic printer) and interfacing to establish connection to large computers.

Software is of central importance. All software should be constructed in a modular form and give access to tuning, measurement and control features by clear menu structured man-machine interfacing. It should allow easy handling of complex data streams. Presentation of measured data should include several modes:

- graphic output of spectra, difference spectra, selected mass intensity versus time plots, concentration and biological activities versus time plots.
- printed output of tables of spectra, single ion measurements, concentrations and biological activities.
- clear analysis and presentation of error conditions with suggested methods to remove sources of these errors.

6 MS Results of Bioprocess Monitoring

The literature on MS applications in bioprocess monitoring is rapidly increasing. The most established mode of MS application in industry is gas analysis for gas balancing.

6.1 Gas Balancing

6.1.1 Estimation of Gas Reaction Rates

The basis for gas balancing is a reliable analysis of gas partial pressures and flow rates. Flow rates of dry gas can reliably be measured using mass flow meters which usually depend only very slightly on pressure and temperature. Gas composition, however, can considerably influence flow rate readings. Figure 16 schematically shows a reactor surrounded by the necessary equipment to make gas balances.

We can set up general balance equations for the liquid phase

$$V_l \frac{dc_{l,i}}{dt} = k_l a(c^*_{l,i} - c_{l,i}) V_l + r_i X V_l \tag{12}$$

and for the gas phase

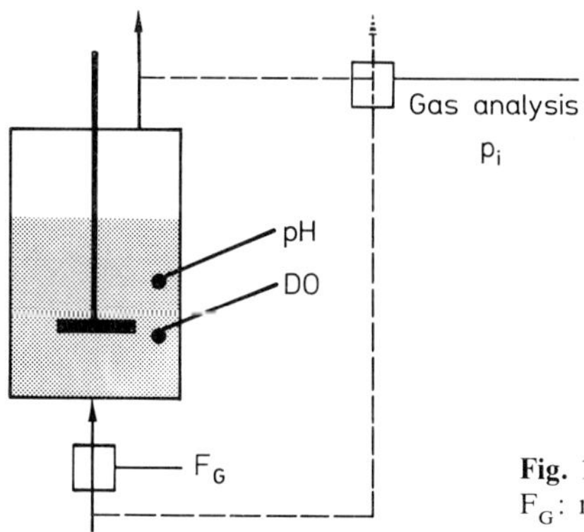

Fig. 16. Bioreactor with equipment to make gas balances. F_G: measurement of gas flow. p_i: Partial pressures

$$V_g \frac{dc_{g,out}}{dt} = F_{in}x_{in,i} - F_{out}x_{out,i} - k_l a(c^*_{l,i} - c_{l,i}) V_l . \tag{13}$$

The meaning of the symbols is: V: volume, c: concentration, $k_l a$: volumetric mass transfer coefficient, c*: equilibrium concentration, r: reaction rate, X: biomass concentration, x: mole fraction, F: flow rate, H: Henry coefficient. The indices are: l: liquid phase, i: component, g: gas phase, in: entrance to the reactor, out: exit of the reactor. The link between gas and liquid phases is given by

$$C^*_{l,i} = x_{out,i} H_i p_{tot} \tag{14}$$

if the gas phase is well mixed.

The most frequent case would be aerobic processes where the inlet gas stream is dry air whose composition is well known and very constant. Argon (0.95%) and usually also nitrogen (78.08%) will be inert gases. Under quasi steady-state conditions where the accumulation term (dc/dt) is very small the inert gas balance will reduce to

$$0 = F_{in}x_{in,inert} - F_{out}x_{out,inert} . \tag{15}$$

From this we can get an estimate of the exit mass flow from measuring the inlet flow F_{in} and inert gas partial pressures

$$F_{out} = F_{in}x_{in,inert}/x_{out,inert} . \tag{16}$$

Combining Eqs. (12), (13), and (16) we get

$$r_i X V_l = -F_{in}(x_{in,i} - x_{out,i}x_{in,inert}/x_{out,inert}) . \tag{17}$$

For oxygen the assumption of quasi steady-state usually will be fulfilled and we can estimate the oxygen uptake rate (OUR)

$$OUR = r_i X V_l = -F_{in}(x_{in,i} - x_{out,i}x_{in,inert}/x_{out,inert}) . \tag{18}$$

For CO_2 the situation often will be more complex mainly because of its dissociation

$$CO_2 + H_2O \rightleftharpoons HCO_3^- + H^+ . \tag{19}$$

Only at low pH values when the concentration of HCO_3^- may be negligible can we neglect balancing liquid phase reactions. The equilibrium constant

$$K = \frac{c_{HCO_3-} c_{H+}}{c_{l, CO_2}} \tag{20}$$

has a value of $K_{30\,°C} = 4.71 * 10^{-7}$ mol L^{-1}. Assuming a well mixed gas phase we can estimate the amount of CO_2 accumulated as HCO_3^- from Eq. (20):

$$c_{HCO_3-} = \frac{K c_{l, CO_2}}{c_{H+}} = \frac{K x_{out, CO_2} p_{tot} H_{CO_2}}{c_{H+}} . \tag{21}$$

Dynamical errors can be significant if pH control is not good enough, because variations in pH will cause accumulation of carbon as bicarbonate. A pH increase of 0.2 units causes a 26% increase in the HCO_3^- concentration if the gas phase concentration of CO_2 is kept constant. The higher the pH the more important this effect will be. At low pH values (<6) and good pH control, we can calculate the CO_2 production rate (CPR) according to Eq. (17) obtaining

$$CPR = r_{CO_2} X V_l = -F_{in}(x_{in, CO_2} - x_{out, CO_2} x_{in, inert}/x_{out, inert}) . \tag{22}$$

In air x_{in, CO_2} is usually very small leading to

$$CPR = F_{in} x_{out, CO_2} x_{in, inert}/x_{out, inert} . \tag{23}$$

In the more general case, where only gas phase accumulation is negligible Eq. (12) has to be modified to include bicarbonate concentration

$$V_l \frac{d(c_{l, CO_2} + c_{HCO_3-})}{dt} = CTR - CPR \tag{24}$$

where CTR is the CO_2 transfer rate which is defined by

$$CTR = F_{out} x_{out, CO_2} - F_{in} x_{in, CO_2} . \tag{25}$$

Combining Eqs. (24), (25), and (16) and introducing the difference quotient instead of the differential quotient we get

$$CPR = F_{in} \left(x_{out, CO_2} \frac{x_{in, inert}}{x_{out, inert}} - x_{in, CO_2} \right) + \frac{\Delta(c_{l, CO_2} + c_{HCO_3-})}{\Delta t} . \tag{26}$$

As c_{1,CO_2} and c_{HCO_3-} are not usually measured on-line, estimates have to be derived from gas analysis and pH measurement. The assumption of equilibrium conditions ($c_{1,CO_2} \approx c_{1,CO_2}{}^*$) between gas and liquid phases will give an estimate of c_{1,CO_2} according to Eq. (13) and of c_{HCO_3-} using Eq (19) giving

$$c_{1,CO_2} + c_{HCO_3-} = x_{out,CO_2} H_{CO_2}(1 + K/10^{-pH}) . \tag{27}$$

For non equilibrium conditions but still negligible gas phase CO_2 accumulation ($dc_g/dt \approx 0$) we get using Eq. (13)

$$(k_1 a)_{CO_2}(c^*_{1,CO_2} - c_{1,CO_2}) V_1 = F_{in} x_{out,CO_2} x_{in,inert}/x_{out,inert} . \tag{28}$$

Assuming $(k_1 a)_{CO_2} \approx (k_1 a)_{O_2}$, which will be reasonable according to Schneider and Frischknecht [104] and Heinzle and Lafferty [48], c_{1,CO_2} and c_{HCO_3-} can be calculated if $(k_1 a)_{O_2}$ is known. This value can either be estimated using empirical correlation from the literature [59] or if c_{1,O_2} can be measured by electrode, balancing for oxygen will give an estimate for $(k_1 a)_{O_2}$ derived from Eq. (13)

$$(k_1 a)_{O_2} = \frac{F_{in}(x_{in,O_2} - x_{out,O_2} x_{in,inert}/x_{out,inert})}{(c^*_{1,O_2} - c_{1,O_2}) V_1} . \tag{29}$$

6.1.2 Error Analysis for Estimation of Gas Reaction Rates

Heinzle et al. [50] made an analysis of the propagation of statistical errors in estimating oxygen uptake rates. Small differences between inlet and outlet oxygen concentrations will be the main reason for amplification of measurement errors. Assuming that inlet concentrations are well known (air) and that gas flow measurement is accurate (mass flow meter) they finally obtained the equation

$$s'_f = ds'_d/(c - d) \tag{30}$$

Table 6. Imprecision of determination of OUR (d) according to Eqs (30) and (34)

$\Delta c'_{O_2}$ [%]	50	10	5	1	Condition
s'_d [%]	4.4	0.67	0.32	0.13	$s'_f = 5\%$
s'_f [%]	1.1	3.5	16	40	$s'_d = 1\%$
$\Delta d'$ [%]	3.3	0.5	0.25	0.049	$\Delta f' = 5\%$
$\Delta f'$ [%]	1.5	9.5	19.4	98	$\Delta d' = 1\%$

$\Delta c'_{O_2}$ — relative concentration difference between inlet and outlet gas concentration;
s'_d — relative statistical error in measuring the ratio of oxygen to nitrogen concentration;
s'_f — relative statistical error in determination of oxygen uptake rate;
$\Delta d'$, $\Delta f'$ — corresponding systematic errors

where s'_f is the relative error for OUR, $d = x_{out,O_2}/x_{out,inert}$, $c = x_{in,O_2}/x_{in,inert}$ and s'_d is the relative error in measuring d. Thus s'_d will also be equal to the error in the ratio of corresponding ion currents. In Table 6 the relation between s'_d and s'_f is illustrated for a series of relative differences in oxygen concentrations of inlet and outlet $\{\Delta c'_{O_2} = 2(x_{in,O_2} - x_{in,O_2})/(x_{in,O_2} + x_{in,O_2})\}$.

Systematic errors in oxygen balancing may be treated as follows. Starting from Eq. (18) we get the correct value

$$\frac{-OUR}{F_{in}x_{in,inert}} = \frac{x_{in,O_2}}{x_{in,inert}} - \frac{x_{out,O_2}}{x_{out,inert}} \tag{31}$$

and substituting for $f = -OUR/F_{in}x_{in,inert}$, $c = x_{in,O_2}/x_{in,inert}$, and $d = x_{out,O_2}/x_{out,inert}$ we get

$$f = c - d\,. \tag{32}$$

If we have a deviation Δd in d we define a relative error $\Delta d' = \Delta d/d$ and $\Delta f' = \Delta f/f$ and get

$$\Delta f' = \Delta d'/(d/c - 1)\,. \tag{33}$$

If OUR and CPR have the same value and if the outlet gas is dry we can simplifiy Eq. (33) to

$$\Delta f' = \Delta d'/(x_{in,O_2}/x_{out,O_2} - 1)\,. \tag{34}$$

Some example values are listed in Table 6. From this table it can be seen that small relative concentration differences ($\Delta c'_{O_2}$) will require very high precision in determination of ratios of ion currents (low values of s'_d and $\Delta d'$). If s'_d or $\Delta d'$ can be kept below 0.1% (as was claimed to be realistic [15,50]) OUR can be determined with sufficient accuracy (small enough values of s'_f and $\Delta f'$) in almost any realistic case. It seems that systematic errors due to inaccurate calibration procedure, lack of sufficiently precise gas mixtures or slight drifsts in the instrument signals, may be more significant than statistical errors. The latter can be reduced by increased measurement time. Drifts seem to be negligible with especially designed magnetic sector instruments [15]. Especially with small quadrupole instruments, the stability will be worse because of the narrow peak form at reasonable resolution values, but repeated calibration with air seems to be sufficient to solve most problems [44].

6.1.3 Gas Balancing with Small Gas Streams

If the amount of gas available for analysis is very small (< 1 ml s^{-1}) as may be the case in laboratory experiments with small volume bacterial cultures or with plant or animal cell cultures, gas analysis can be made using membrane probes. This method has been used by Pungor et al. [89] for bakers' yeast culture.

6.1.4 Examples

Compared to its importance and general applicability there are only few examples of gas balancing work using MS in the literature. Generally it can be expected that this method is independent of the type of biological culture. The problems which may be expected in practice will be similar in most cases.

It is well known that a number of laboratories and pilot plants are now equipped with on-line MS gas analyzers. There are, however, only few reports from industry [7,8,9,15,16,118]. MS was used in a number of fermentations including those based on bacterial and fungal processes. The method was reported to be very reliable. The only disadvantages reported were the rather high investment cost and rare but expensive maintenance. According to private communications from a pharmaceutical company there seem to be problems with long term exposure to antifoam aerosols.

Pungor et al. [89,92] describe balancing of bakers' yeast cultures. Furukawa et al. [38] and Heinzle et al. [45] analyzed continuous yeast culture under steady-state and oscillatory conditions. Heinzle et al. [50] monitored *Bacillus subtilis* culture. Heinzle and Dettwyler [44] applied a small residual gas quadrupole MS for gas balancing and further elemental balancing of an *Alcaligenes latus* culture to monitor biomass, the product poly-β-hydroxybutyric acid (PHB) and substrates NH_4^+ and sugar. In this case error propagation from either errors in measurement of gas partial pressures (especially for oxygen) or of pH was found to be critical because of rather small differences in the oxidation state of carbon in cells and in the product PHB.

6.2 Dissolved Gases and Volatiles

6.2.1 Dissolved Gases

Hoch and Kok in 1963 pioneered the application of an MS membrane probe to monitor dynamics of a biochemical system [53]. They analyzed dissolved oxygen and carbon dioxide in algal photosynthetic suspensions. Monitoring of dissolved gases in blood using MS membrane probes was described in 1966 by Woldring et al. [137]. The method was applied to sterile biotechnological processes a couple of years later [96]. Lundsgaard et al [71] measured oxygen uptake rates in biochemical systems. Physiological studies monitoring dissolved O_2 and CO_2 were carried out by Ponte and Purves [87]. Jouanneau et al. [60] and Heinzle and Lafferty [48] described monitoring of dissolved hydrogen in bacterial cultures (Fig. 17).

Pungor et al. [89] monitored dissolved O_2 and CO_2 in a bakers' yeast batch culture. As they used an external measurement cell, monitoring of dissolved O_2 may have been limited due to possible mass transfer and reaction within the measurement loop. Jensen et al. [56] measured dissolved N_2, H_2, and O_2 in bacterial nitrogen fixation studies. Heinzle et al [45] further evaluated measurement of dissolved O_2 and CO_2 in yeast culture. Lloyd et al. [70] studied the Pasteur effect and the effect of NH_4^+ [146] in yeast, measuring dissolved O_2 and CO_2. Lloyd's group at University College, Cardiff, studied dynamics of anaerobic systems measuring dissolved H_2, CH_4, and O_2 [107,144,145,148–155]. Joergensen [57] studied the mechanism of methane monooxygenase in methanotrophic bacteria.

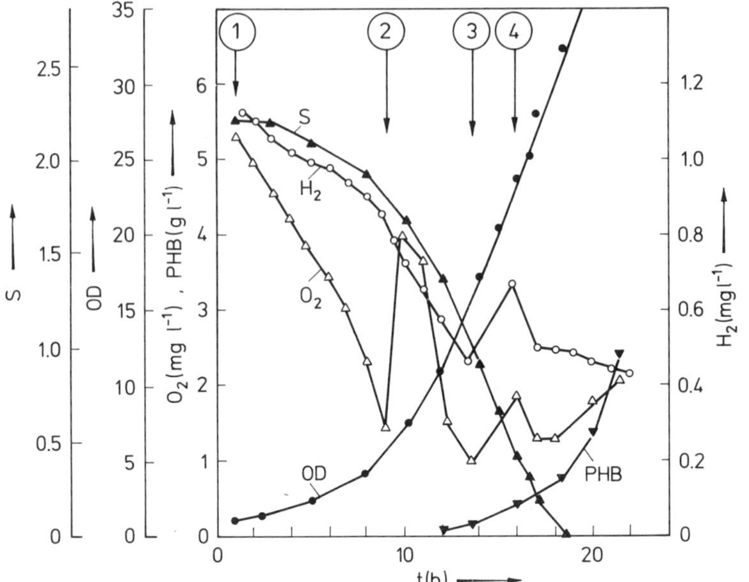

Fig. 17. Monitoring of dissolved H_2 and O_2 in *Alcaligenes eutrophus* batch culture [48] ▲, s: substrate; ●, OD: optical density; △, O_2: dissolved oxygen; ○, H_2: dissolved hydrogen; ▼, PHB: poly-β-hydroxy-butyric acid; 1 to 4 indicate changes in the gas supply:
① start of culture 980 ml min^{-1}; 80% H_2, 15% O_2, and 5% CO_2,
② 1,050 ml min^{-1}; 76% H_2, 17% O_2, and 7% CO_2,
③ 1,580 ml min^{-1}; 75% H_2, 16% O_2, and 9% CO_2,
④ addition of polypropylenglycol to reduce foaming (0.1 ml per 10 l culture volume)

To study microbial pathways isotope studies using stable non-radioactive isotopes ^{18}O and ^{13}C were carried out by Hoch and Kok [53], Radmer and Ollinger [93] and by Silverman [110].

Degn et al. [31], Lloyd et al. [68,147,156], and Bohátka et al. [8] recently produced extensive reviews on continuous measurement of dissolved gases in biochemical systems using MS membrane probes. Generally there seems to be a straightforward approach to monitor dissolved gases. Quantitative measurements, however, may be disturbed by the presence of other gases [32].

6.2.2 Monitoring of Volatiles

Besides gases, a number of volatiles are of considerable interest in biotechnological processes (ethanol, methanol, acetone, butanol, acetic acid, acetoin, butanediol, flavour compounds, etc.). These compounds can usually be easily measured using GC. New developments in wide-bore capillary column systems using glass liners even allow injection of total reaction mixture for direct and fast analysis of volatiles [41]. Sampling is still necessary to transfer samples from the bioreactor to the GC. Another approach which may be automatized much more easily is head space analysis using capillary GC [23].

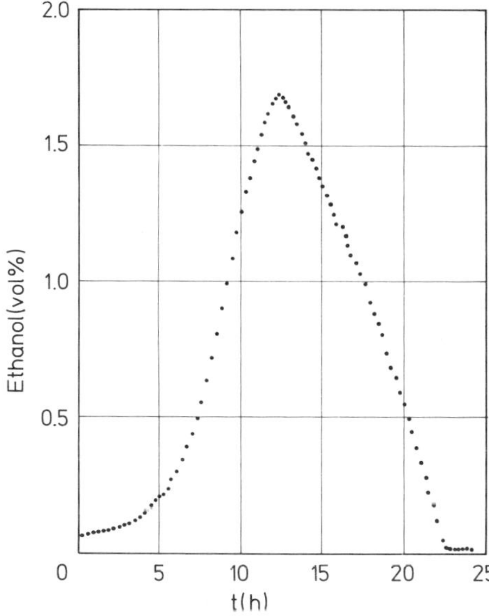

Fig. 18. Monitoring of ethanol in bakers' yeast culture [92]

MS membrane probes allow continuous measurement of volatiles. Because of the different chemical affinities of volatiles for the membrane used, sensitivity and dynamic response may vary considerably.

Reuss et al. [96] monitored dissolved methanol concentration as well as dissolved gases. Pungor et al. [89], as can be seen in Fig. 18, monitored ethanol concentration during yeast batch culture: diauxic growth is evident.

Calvo et al. [17,18] monitored ethyl-2-furoate to study kinetics of enzymatic transesterification of p-nitrophenyl-2-furoate using α-chymotrypsin.

Heinzle et al. [45,46] obtained detailed information on the dynamic behaviour of continuous culture of bakers' yeast and this enabled them to devise a mathematical model for such cultures. Fig. 19 shows a culture under oscillatory and steady-state conditions. The supply of oxygen to the culture was varied as indicated by arrows which influenced the occurrence and frequency of oscillation. In this example both gas and liquid phases were analyzed using one single MS with capillary and membrane inlet.

Propanol was used as a carbon source in erythromycin production which was continuously monitored [7,8,9].

Doerner et al. [32] reported considerable problems in the analysis of acetone-butanol production. Obviously under these conditions (with rather high concentration of volatile organic compounds) pure compound calibration was not sufficient because of the non-linear behaviour of the membrane transport process. Multi-component calibration would allow quantitative on-line analysis. This rather time-consuming procedure would, however, considerably diminish the potential advantages of MS for the monitoring of volatiles. Fig. 20 shows a comparison between GC results and MS measurements with pure component calibration for n-butanol. The deviation becomes more and more significant towards the end of growth, where concentrations of volatiles became higher.

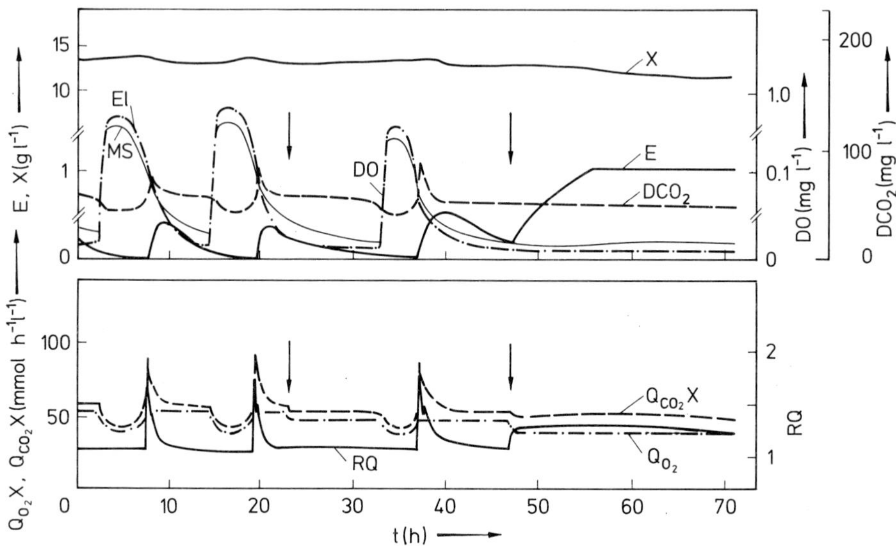

Fig. 19. Oscillatory continuous bakers' yeast culture [45]
Upper part: liquid phase analysis. E: ethanol; X: biomass (OD), DO: dissolved oxygen; El: electrode; DCO_2: dissolved CO_2. Lower part: gas balancing results. $Q_{02}X$: oxygen uptake rate; QCO_2X: CO_2 production rate. Arrows indicate changes in oxygen supply

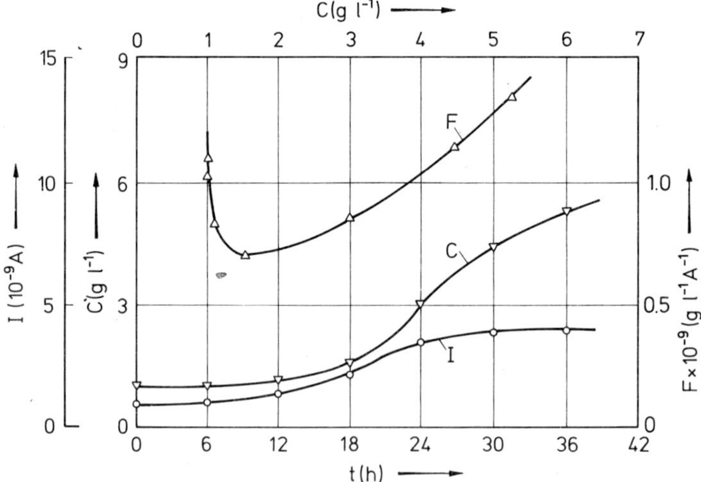

Fig. 20. Comparison of GC and MS measurement in acetone-butanol production [32] FC: concentration of n-butanol (GC); I: ion current F: Calibration factor = concentration (GC)/ion current.

Griot et al. [41] monitored acetoin and butanediol in *Bacillus subtilis* culture (Fig. 21). Optimization of vacuum tubing material and temperature very much improved measurement performance. Medium temperatures (≈150 °C) and Teflon tubing were found to be optimal. Response time was much shorter using a 75 μm silicone rubber membrane consisting of three layers of 25 μm each; this precaution

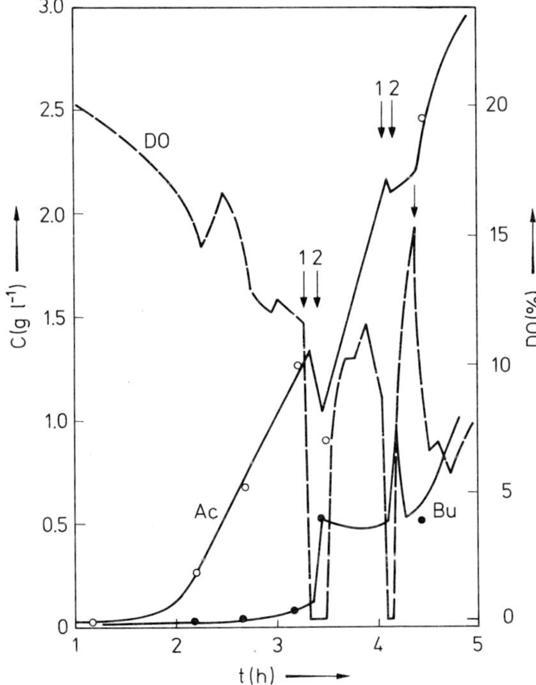

Fig. 21. Comparison of GC and MS measurement in *Bacillus subtilis* batch culture[41] ● butanediol (GC); full lines show MS results from single component calibration; DO: dissolved oxygen. (1) end and (2) start of aeration

reduced the probability of holes in the final membrane. In this case calibration with pure compounds was sufficient as comparisons with GC analysis confirmed. Acetoin was measured using the pure mass 86. Butandediol mainly contributed to the intensity of mass 57. Estimation of concentrations followed the following simple procedure

$$c_{Ac} = k_{86,Ac} I_{86} \tag{35}$$

$$c_{Bu} = (I_{57} - c_{Ac}/k_{57,Ac}) \, k_{57,Bu} \, . \tag{36}$$

Pungor et al.[90] used a membrane interface to monitor concentration of ^{2}HHO in a H_2O solution. This stable isotope may be used for non-radioactive physiological studies.

6.2.3 Fingerprinting

In many cases product concentrations or production capacity of biological cultures are very difficult to analyze and usually cannot be measured on-line. In some special cases it has been possible to monitor biomass[78,127], biomass and gluconic acid[157] or poly-β-hydroxybutyric acid[44] using balancing methods based on gas analysis results. In most bioprocesses with secondary metabolite production, stoichiometry is not known or variable because of unknown effects. The product concentration is often very small compared to other substances in the medium. Therefore balancing methods cannot be applied easily.

There is, however, still another approach for the collection of information about the bioprocess. Indirect methods like monitoring redox-potential [4,62] or measurement of filtration characteristics [82] can give information of culture activity.

A somewhat related approach, namely monitoring of mass spectra of volatiles in culture media was suggested by Heinzle et al. [47]. They found that in a number of industrially important microbial cultures including *Clostridium tetanii*, *Corynebacterium diphtheriae*, several *Streptomyces* species, *Bordetella pertussis*, *E. coli* and *Actinomyces* species, characteristic spectra of volatiles can be detected using mass spectrometer membrane probe.

A spectrum from a *Clostridium tetanii* culture is shown in Fig. 22.

It can be seen that the water peaks (m/z = 18 and m/z = 17) and peaks from dissolved gases (nitrogen: m/z = 28; carbon dioxide: m/z = 44; oxygen: m/z = 32) dominate. Each process type examined gave characteristic spectra which differed considerably. It was observed that spectra vary during cultivation. To further analyze sets of spectra from sets of cultures, factor analysis was applied. Results from three seperate experiments with a *Streptomyces* species together with off-line measurements of the actual product concentrations are given in Fig. 23.

Batches B3 and B4 gave comparable product formation results, whereas batch B6 had a much lower product yield. From factor analysis it might be concluded that a different medium or inoculum was used in batch B6, because the initial factor value differs significantly from the other two batches. It was found that product formation and the factor score of factor 1 can be correlated. From each factor value (f_1), the corresponding initial value ($f_{1,0}$) was subtracted and plotted versus product concentration. The linear correlation found could, however, not be used to exactly describe the time course of the product concentration for the individual experiments.

Finally a methodology was suggested to apply fingerprinting to any kind of

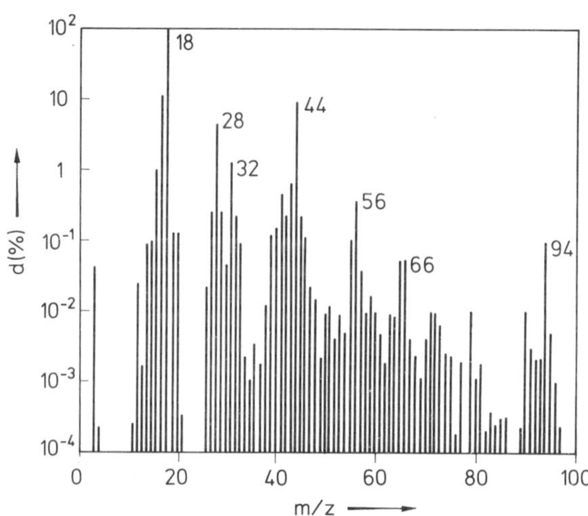

Fig. 22. Spectrum from *Clostridium tetanii* fermentation [47]: d: intensity relative to I_{18}; m/z-mass units. Here, $I_{18} = I_{max} = 1.29 * 10^{-6}$ A

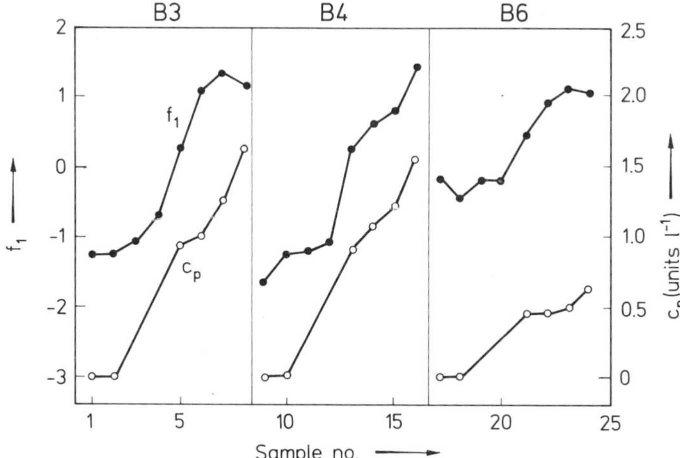

Fig. 23. Comparison of product concentration with factor 1 (f_1) in three different batch cultivations of *Streptomyces* species [47]. Symbols are: (○) product concentration (c_p) [u l^{-1}] and (●) f_1

process where direct on-line measurement of product concentration or culture activity is not possible. In these cases indirect measurements can improve process analysis and control. After checking for the existence of volatiles detectable by the MS membrane probe, a product factor correlation may be established. If a useful correlation can be found the time course of product formation can be easily monitored even with a small microcomputer. Occasional updating of the product factor correlation was suggested. This would be especially necessary if considerable changes in the composition of the medium are made.

6.3 Further Methods

6.3.1 Increasing Volatility by Chemical Reaction

Some of the biological substrates or reaction products are not volatile under process conditions at physiological pH. Such volatile acids or bases may be analyzed after sampling and proper adjustment of pH to increase volatility [131]. Such a variable pH interface allows measurement of ammonia, bicarbonate, aliphatic volatile carbonic acids, pyruvic acid and others. An interface of this kind was used by Bohátka et al. [8] to measure ammonia in biological media. The pH was increased by addition of NaOH. Bohátka [7] also used the same interface to analyse total carbonate by addition of acid and measurement of liberated CO_2. Heinzle and Dettwyler [44] determined dissolved CO_2 and HCO_3^- during cultivation of *Alcaligenes latus*.

Detection limits for determination of acids and bases may vary considerably depending on volatility and membrane permeability [131]. Heinzle et al. [45] reported increasing sensitivity in the measurement of carbonic acids with increasing length of the aliphatic chain.

A further extension for the application of MS membrane probes was described by Pungor et al. [91]. They monitored concentration of 2-oxo-glutaric acid in penicillin cultivation by continuous esterification with methanol. A correlation between concentration of 2-oxo-glutarate and peak intensity was found. The most significant limitation was identified to be the slow reaction rate.

6.3.2 Pyrolysis

Pyrolysis MS is a well established method [76] in particular for characterizing complex polymer materials. In a non-oxidative environment, most organic materials give characteristic volatile reaction products if they are heated to 300–600 °C in a reproducible manner. The most successful method seems to be Curie-point pyrolysis in the vacuum of the MS. This method guarantees quick heating to well defined temperatures. The most significant problem in applying this method to on-line analysis of bioprocesses is probably the fact that very small amounts (about 100 μg) of sample have to be put onto a thin wire. This would make it very difficult to get representative samples from media. Automatization of sampling does not seem to be straightforward in this case.

Windig et al. [136] applied Curie-point pyrolysis MS to the analysis of polymer mixtures. Data were analyzed using factor analysis. In another paper they described the chemical characterization of yeast-like fungi using similar methods [135]. Anhalt and Fenselau [2] suggested the use of pyrolysis as a means to identify bacteria.

Recently Heinzle et al. [49] attempted an analysis of poly-β-hydroxybutyric acid (PHB) in a growth of *Alcaligenes latus* using a simple oven to pyrolyse the whole reaction mixture without any pretreatment. Heating up of the sample in a He atmosphere was rather slow. During the heating period, a maximum intensity of the characteristic factor spectra could be identified. These maxima were used for correlation with PHB concentration. Despite this crude pyrolysis method, additions of variable amounts of PHB to identical bacterial samples gave very good linear correlation with factor 4. Application of this method to a batch process showed that some significant disturbances can be observed (Fig. 24).

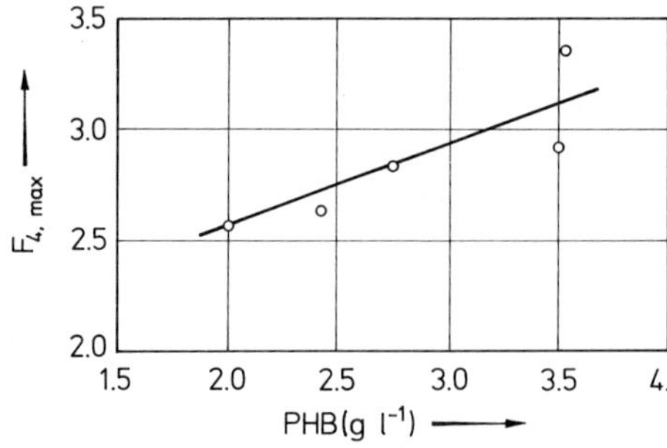

Fig. 24. Correlation between maximum of factor 4 from pyrolysis MS spectra ($F_{4,max}$) and PHB concentration.

7 Future Trends

In gas analysis, reliability and accuracy may well become further improved. This will be especially important when using elemental balancing methods to monitor process variables indirectly. For liquid-phase analysis of dissolved gases (and especially volatiles), the introduction of new membrane materials may lead to improvements in selectivity, accuracy, and speed of analyses. Multivariate statistical methods may give interesting results in industrial process analysis of volatiles.

If it becomes possible to design a reliable method to reproducibly and automatically pyrolyse reaction samples, pyrolysis-MS could provide a very valuable contribution to process analysis of biotechnological processes. It is not, however, expected that this method will be easily applied to any process.

A number of interesting developments stem from the introduction of new or improved ionization methods, especially of soft ionization techniques. These may simplify the analysis of mixtures of volatiles because of the resulting simplified fragmentation patterns. Recently, a MS instrument using chemical ionization designed for process applications has been announced [143)]. For the analysis of non-volatile compounds, some developments in HPLC-MS coupling may bring a real breakthrough. The most promising method at present is thermospray interfacing and ionization [122,125)]. In conventional methods non-volatile salts still create considerable problems for long term operation [39)].

8 Concluding Remarks

Application of MS to biotechnological process analysis greatly increases the number of compounds that can be monitored on-line. Gas analysis of gases of interest is possible using one single instrument. The use of internal standard methods helps to improve reproducibility and accuracy.

Additional monitoring of all dissolved gases of interest and analysis of a number of volatiles has been shown to be possible even with a sterilizable membrane probe. Transport process of compounds to the membrane is important when reactions occur within the liquid boundary layer or when using thin membranes with high permeability. Non-linear behaviour of membrane penetration processes may create problems at high concentrations of volatiles. For a number of compounds connection tube material to, and temperature of the ion source have to be carefully selected. Fingerprinting of volatiles may be applicable to a number of industrial processes where process analysis is very difficult.

9 Symbols and Abbreviations

a	surface area	$[m^2]$
c	concentration	$[mol\ L^{-1}]$, $[g\ L^{-1}]$
c	number of measurements	
C	concentration matrix	
CUR	carbon dioxide uptake rate	$[mol\ s^{-1}]$
CTR	carbon dioxide transfer rate	$[mol\ s^{-1}]$
d	deviation	$[s\ m^{-1}]$
D	diffusion coefficients	$[m^2\ s^{-1}]$
DO	dissolved oxygen concentration	$[g\ L^{-1}]$

F	flow rate	[mol s^{-1}]
H	Henry coefficient	[mol m^{-3} Pa^{-1}]
I	ion current	[A]
I	matrix of ion currents	
K	equilibrium constant	[mol L^{-1}]
k	MS calibration constant	[mol m^3 A^{-1}]
k_1a	volumetric mass transfer coefficient	[s^{-1}]
l	membrane thickness	[m]
M	molecular weight	[g mol^{-1}]
n	number of components	
OD	optical density	
OUR	oxygen uptake rate	[mol s^{-1}]
OTR	oxygen transfer rate	[mol s^{-1}]
p	pressure	[Pa]
r	reaction rate	[mol L^{-1} s^{-1}]
r	number of masses	
R	gas constant	[Pa m^3 K]
R	film resistance	[s m^{-1}]
RQ	respiratory quotient	
s	sensitivity coefficient	[A l g^{-1}]
S	matrix of sensitivity coefficients	
t	time	[s], [h]
T	temperature	[K], [°C]
V	volume	[m^3]
x	mole fraction	
X	biomass concentration	[g m^{-3}]
τ	residence time, response time	[s]

Subscripts and Superscripts:

b	refers to boundary layer
g	refers to gas phase
in	refers to reactor inlet
l	refers to liquid phase
m	refers to membrane
out	refers to reactor outlet
tot	total
*	refers to equilibrium value
′	relative

Abbreviations:

DC	direct current
FAB	fast atom bombardment ionization
FD	field desorption ionization
FE	field emission ionization
FT	Fourier transform
EI	electron impact ionization
GC	gas chromatography

HPLC high performance liquid chromatography
i.d. inner diameter
MS mass spectrometer (-metry)
m/z mass to charge ratio
SEM secondary electron multiplier
SIM selected ion monitoring

10 References

1. Alberti, J. C., Phillips, J. A., Fink, D. J.: Off-line monitoring of fermentation samples by FTIR/ATR: a feasibility study for real-time process control. 7th Symp. on "Biotechnology of Fuels and Chemicals, Gatlinburg", Tennessee, May 14–17, 1985
2. Anhalt, J. P., Fenselau, C.: Anal. Chem. *47*, 21 (1975)
3. Batey, J. H.: Int. J. Mass Spectrom. Ion Phys. *60*, 117 (1984)
4. Berovic, M., Cimerman, A.: Eur. J. Appl. Microbiol. Biotechnol. *16*, 165 (1982)
5. Beyeler, W., Einsele, A., Fiechter, A.: ibid. *13*, 10 (1981)
6. Boelcke, C., Lenz, R., Peckmann, U., Reuss, M.: Reaktorgekoppelte Filtrationsanlage mit automatischem Probeneinlaß zur Aufbereitung und Analyse von Proben aus Fermentationsprozessen (Bio-Filtrator). Technical Report TU Berlin 1985
7. Bohátka, S.: Quadrupole mass spectrometric measurement of dissolved and free gases. In: "Gas Enzymology" (Degn, H., et al., eds.), p. 1–16, Reidel Publ. Company 1985
8. Bohátka, S., Kálmán, P., Langer, G., Szilágyi, J.: Quadrupole mass spectrometer analyser for monitoring fermentation processes. In: "Advances in Fermentation", p. 131, Wheatland Journals Ltd 1983
9. Bohátka, S. Langer, G., Szilágyi, J., Berecz, I.: Int. J. Mass Spectrom. Ion Phys. *48*, 277 (1983)
10. Boon, J. J., Tom, A., Brandt, B., Eijkel, G. B., Kistemaker, P. G., Notten F. J. W., Mikx, F. H. M.: Anal. Chim. Acta *163*, 193 (1984)
11. Brantigan, J. W.. Gott, V. L., Vestal, M. L., Fergusson, G. J., Johnson, W. H.: J. Appl. Physiol. *28*, 375 (1970)
12. Breth, A., Dobrozemsky, R., Kraus, B.: Vacuum *33*, 73 (1983)
13. Brockman, T. J., Anderson, L. B.: Anal Chem. *56*, 207 (1984)
14. Brunnee, C., Voshage, H.: „Massenspektrometrie". K. Thieme, Muenchen 1968
15. Buckland, B., Brix, T., Fastert, H., Gbewonyo, K., Hunt, G., Deepak, J.: Bio/technology *3*, 982 (1985)
16. Buckland, B., Fastert, H.: Analysis of fermentation exhaust gas using a mass-spectrometer. In: "Computer applications in Fermentation Technology", p. 119, Soc. of Chemical Industry, London 1982
17. Calvo, K. C., Weisenberger, C. R., Anderson, L. B., Klapper, M. H.: Anal. Chem. *53*, 981 (1981)
18. Calvo, K. C., Weisenberger, C. R., Anderson, L. B., Klapper, M. H.: J. Am. Chem. Soc. *105*, 6935 (1983)
19. Cameresi, G. G., Costa, B.: Adv. Mass Spectrom. *7B*, 1062 (1978)
20. Chapman, J. R.. Computers in mass spectrometry. Academic Press, London 1978
21. Cleland, N., Enfors, S. O.: Anal. Chim. Acta *163*, 281 (1984)
22. Cleland, N., Hörnstein, E. G., Elwing, H., Enfors, S. O., Lundström, I: Appl. Microbiol. Biotechnol. *20*, 268 (1984)
23. Comberbach, D. M., Bu'lock, J. D.: Biotechnol. Bioeng. *25*, 2503 (1983)
24. Cox, R. P., Jensen, B. B., Joergensen, L., Degn, H.: Microbiological Sciences *1*, 200 (1984)
25. Crawford, R. W., Bedford, R. G., Wong, C. M., Brand, H. R., Kishiyama, K. I.: Rev. Scient. Instrum. *20*, 333 (1981)
26. Crawford, R. W., Bedford, R. G., Wong, C. M., Brand, H. R., Kishiyama, K. I.: Dyn. Mass Spectrometry *6*, 195 (1981)
27. Dairaku, K., Yamané, T.: Biotechnol. Bioeng. *21*, 1671 (1979)

28. Danielsson, B., Mandenius, C. F., Winquist, F., Mattiasson, B., Mosbach, K.: Fermenter monitoring and control by enzyme thermistor. In: Advances in Biotechnology *1*, p. 445 (Moo-Young, M., Robinson, C. W., Vezina, C., eds.), Pergamon Press, Toronto 1981
29. Dekkers, R. M.: State estimation of a fed-batch baker's yeast fermentation. In: "Modelling and Control of Biotechnical Processes", (Halme, A., ed.), Pergamon Press, Oxford 1983
30. Dekkers, R. M., Voetter, M. H.: Adaptive control of a fed-batch baker's yeast fermentation. In: "Modelling and Control of Biotechnological Processes", p. 73 (Johnson, A., ed.), Pergamon Press, Oxford 1985
31. Degn. H., Cox, R. P., Lloyd, D.: Methods Biochem. Anal. *31*, 165 (1985)
32. Doerner, P., Lehmann, J. Piehl, H., Megnet, R.: Biotechnol. Letters *4*, 557 (1982)
33. Endo, I., Nagamune, T., Nakamura, T., Inoue, I.: A successive fed-batch culture of a brewer's yeast in three substrates system. In: "Third European Congress of Biotechnology" Vol. *2*, p. 437 (DECHEMA ed.), Verlag Chemie, Weinheim 1984
34. Enfors, S. O., Cleland, N.: Calibration of oxygen- and pH-based enzyme electrodes for fermentation control. In: "Chemical Sensors", p. 672 (Seiyama, T., ed.), Elsevier, Amsterdam 1983
35. Eustache, H., Histi, G.: J. Membrane Sci. *8*, 105 (1981)
36. Finnigan MAT.: Modell ITD 700, 1985
37. Fite, W. L., Patterson, T. A., Siegel, M. W., Brackmann, R. T.: Dyn. Mass Spectrometry *6*, 167 (1981)
38. Furukawa, K., Heinzle, E., Dunn, I. J.: Biotechnol. Bioeng. *25*, 2293 (1983)
39. Goetz, P.: Kopplung von Massenspektrometer und Bioreaktor: Entwicklung und Erprobung der Thermospray-Einlaßtechnik. Diplomarbeit, TU Berlin 1985
40. Graves, D. J., Baumgardner, J. E., Neufeld, G. R., Quinn, J. A.: AIChE Symp. Ser. *79*, 124 (1983)
41. Griot, M., Heinzle, E., Dunn, I. J.: Fast determination of volatiles in fermentation broth using capillary GC and mass spectrometry, to be published
42. Grosz, R., Stephanopoulos, G., San, K. Y.: Biotechnol. Bioeng. *26*, 1198 (1984)
43. Heinzle, E., Bolzern, O., Dunn, I. J., Bourne, J. R.: A porous membrane-carrier gas measurement system for dissolved gases and volatiles in fermentation systems. In: "Advances in Biotechnology" *1*, p. 439 (Moo-Young, M., Robinson, C. W., Vezina, C., eds.), Pergamon Press, Toronto 1981
44. Heinzle, E., Dettwyler, B.: Einsatz von Elementbilanzen zur on-line Analyse von Fermentationen. Produktion von Poly-β-Hydroxybuttersäure. 4. DECHEMA Jahrestagung der Biotechnologen, 3/4 Juni, Frankfurt/Main 1986
45. Heinzle, E., Furukawa, K., Dunn, I. J., Bourne, J. R.: Bio/technology *1*, 181 (1983)
46. Heinzle, E., Furukawa, K., Tanner, R., Dunn, I. J.: Modelling of sustained oscillations observed in continuous culture of *Saccharomyces cerevisiae*. In: "Modelling and Control of Biotechnical Processes", p. 57 (Halme, A., ed.), Pergamon Press, Oxford 1983
47. Heinzle, E., Kramer, H., Dunn, I. J.: Biotechnol. Bioeng. *27*, 238 (1985)
48. Heinzle, E., Lafferty, R. M.: Europ. J. Appl. Microbiol. Biotechnol. *11*, 17 (1980)
49. Heinzle, E., Lunden, M., Dunn, I. J.: Analysis of biomass and metabolites using pyrolysis mass spectrometry. In: "Modelling and Control of Biotechnological Processes", p. 61 (Johnson, A., ed.), Pergamon Press, Oxford 1985
50. Heinzle, E., Moes, J., Griot, M., Kramer, H., Dunn, I. J., Bourne, J. R.: Anal. Chim. Acta *163*, 219 (1984)
51. Heinzle, E., Moes, J., Griot, M., Sandmeier, E., Dunn, I. J., Bucher, R.: Ann. N. Y. Acad. Sci. *469*, 178 (1986)
52. Hitchman, M. L.: Measurement of dissolved oxygen, Wiley, New York 1978
53. Hoch, G., Kok, B.: Arch. Biochem. Biophys. *101*, 160 (1963)
54. Hunter, J. A., Stacy, R. W., Hitchcock, F. A.: Rev. Sci. Instrum. *20*, 333 (1949)
55. Hwang, S. T., Kammermeyer K.: Membranes in Separation, J. Wiley & Sons (Wiley Interscience), New York 1975
56. Jensen, B. B., Cox, R. P.: Spec. Publ. Soc. Gen. Microbiol. *14*, 279 (1985)
57. Joergensen, L.: The reaction mechanism of methane monooxygenase studied by membrane-inlet mass spectrometry in whole cells of methanotrophic bacteria. In: "Gas Enzymology", p. 187 (Degn, H., et al. eds.), Reidel, Dordrecht 1985

58. Johnson, T. D., Watkins, G. M., Holsinger, J., Roberts, M. P., Thomas, D. D.: Recent Dev. Mass Spectrom. Biochem. Med. *5*, 463 (1979)
59. Joshi, J. B., Pandit, A. B., Sharma, M. M.: Chem. Eng. Sci. *37*, 813 (1982)
60. Jouanneau, Y., Kelly, B. C., Berlier, Y., Lespinat, P. A.: J. Bacteriol. *143*, 628 (1980)
61. Kienitz, H.: Massenspektrometrie, Verlag Chemie, Weinheim/Bergstrasse 1968
62. Kjaergaard, L.: Adv. Biochem. Eng. *7*, 131 (1977)
63. Kroner, K. H., Kula, R. M.: Anal. Chim. Acta *163*, 3 (1984)
64. Langer, G., Berecz, I., Bohátka, S.: Acta Phys. Acad. Sci. Hung. *49*, 307 (1980)
65. Langer, G., Bohátka, S., Berecz, I., Schlenk, B. Boernemisza-Pauspertl, P., Kiss, K., Buzás, I., Pártay, G., Mázsik, L., Sági, F.: Vacuum *34*, 757 (1984)
66. Lenz, R., Boelcke, C., Peckmann, U., Reuss, M.: A new automatic sampling device for the determination of filtration characteristics and the coupling of an HPLC to fermenters. In: "Modelling and Control of Biotechnological Processes", p. 55 (Johnson, A., ed.) Pergamon Press, Oxford 1985
67. Linnarson, A.: Adv. Mass Spectrom. *8 B*, 1559 (1980)
68. Lloyd, D., Bohátka, S., Szilágyi, J.: Biosensors *1*, 179 (1985)
69. Lloyd, D., James, K., Williams, J., Williams, N.: Anal. Biochem. *116*, 17 (1981)
70. Lloyd, D., Scott, R. I.: J. Microbiol. Methods *1*, 313 (1983)
71. Lundsgaard, J. S., Peterson, L. C., Degn, H.: Mass spectrometric determination of oxygen kinetics in biochemical systems. In: "Measurement of Oxygen", p. 163 (Degn, H., Brook, R., eds.), Elsevier, Amsterdam 1976
72. Malinowski, E. R.: Anal. Chem. *49*, 612 (1977)
73. Malinowski, E. R., Howery, D. G.: Factor analysis in chemistry, J. Wiley & Sons, New York 1980
74. Mandenius, C. F., Danielsson, B., Mattiasson, B.: Anal. Chim. Acta *163*, 135 (1984)
75. Matz, G.: Ein mobiles Massenspektrometer zur Erfassung umweltbelastender Schadstoffe. Hochschule der Bundeswehr, Hamburg, Ph. D. Thesis 1981
76. Meuzelaar, L. C., Haverkamp, J. Hileman, F. D.: Pyrolysis mass spectrometry of recent and fossil biomaterials, Elsevier, Amsterdam 1982
77. Millard, B. J.: Quantitative mass spectrometry. Heyden, London 1978
78. Moo, D. G., Cooney, C. L.: Biotechnol. Bioeng. *25*, 225 (1983)
79. Multala, R., Aittamaa, J., Salminen, K. K., Halmu, A., Järveläinen, M.: Computers & Chem. Eng. *3*, 47 (1979)
80. Mütze, B.: Biotechnol. Bioeng. *26*, 390 (1984)
81. Muysers, K., Smidt, U.: Clinical uses of mass spectrometry. In: "Biochemical applications of mass spectrometry", p. 601, (Waller, G. R., ed.), Wiley Interscience, New York 1972
82. Nestaas, E., Wang, D. I. C.: Biotechnol. Bioeng. *25*, 1981 (1983)
83. Norris, P. E., Scrivens, J. H.: Dyn. Mass Spectrometry *6*, 155 (1981)
84. Ottley, T. W.: ibid. *6*, 212 (1981)
85. Pinnick, W. J., Lavine, B. K., Weisenberger, C. R.: Anal. Chem. *52*, 1102 (1980)
86. Poll, A., Potter, C. J., Tily, P. J.: Chem. Eng. London *N 244*, CE413 (1970)
87. Ponte, J., Purves, M. J.: Recent Dev. Mass Spectrom. Biochem. Med. *6*, 483 (1979)
88. Puhar, E., Einsele, A., Bühler, H., Ingold, W.: Biotechnol. Bioeng. *22*, 2411 (1980)
89. Pungor, E., Perley, C. R., Cooncy, C. L., Weaver, J. C.: Biotechnol. Letters *2*, 409 (1981)
90. Pungor, E., Klibanov, A. M., Cooney, C. L., Weaver, J. C.: Biomed. Mass Spec. *9*, 181 (1982)
91. Pungor, E., Pecs, M., Szigeti, L., Nyeste, L., Szilagyi, J.: Anal. Chim. Acta *163*, 185 (1984)
92. Pungor, E., Schaefer, E., Weaver, J. C., Cooney, C. L.: Direct monitoring of a fermentation in a computer mass spectrometer fermentor system. In: Advances in Biotechnology *1*: p. 393 (Moo-Young, M., Robinson, C. W., Vezina, C., eds.), Pergamon Press, Toronto 1981
93. Radmer, R., Ollinger, O.: Methods in Enzymology *69*, 547 (1980)
94. Ramsey, G., Turner, A. P. F., Franklin, A., Higgins, I. J.: Rapid bioelectrochemical methods for the detection of living microorganisms. In: "Modelling and Control of Biotechnological Processess", p. 65 (Johnson, A., ed.), Pergamon Press, Oxford 1985
95. Rautenbach, R., Albrecht, R.: Chem. Ing.-Tech. *54*, 229 (1982)

96. Reuss, M., Piehl, H., Wagner, F.: Europ. J. Appl. Microbiol. Biotechnol. *1*, 323 (1975)
97. Roels, J. A.: Energetics and Kinetics in Biotechnology, Elsevier Biomedical, Amsterdam 1983
98. Rummel, K.: Applied Factor Analysis, North Western University Press, Evanston 1970
99. Ryhage, R.: Anal. Chem. *36*, 759 (1964)
100. San, K. Y., Stephanopoulos, G.: Biotechnol. Bioeng. *26*, 1189 (1984)
101. San, K. Y., Stephanopoulos, G.: ibid. *26*, 1209 (1984)
102. Schlichting, H.: Boundary-layer theory. Mc Graw-Hill, New York 1968
103. Schmidt, W. J., Meyer, H. D., Schügerl, K., Kuhlmann, W., Bellgardt, K. H.: Anal. Chim. Acta *163*, 101 (1984)
104. Schneider, K., Frischknecht, K.: J. Appl. Chem. Biotechnol. *17*, 631 (1977)
105. Schorr, W. K., Duschner, H., Starke, K.: Anal. Chem. *54*, 671 (1982)
106. Schuy, K. D.: Z. Instr. *75*, 190 (1967)
107. Scott, R. I., Williams, T. N., Whitmore, T. N., Lloyd, D.: Eur. J. Appl. Microbiol. Biotechnol. *18*, 236 (1983)
108. Scrivens, J. H.: Vacuum *32*, 169 (1982)
109. Scrivens, J. H., Ramage, J. C.: Int. J. Mass Spectrom. Ion Phys. *60*, 299 (1984)
110. Silverman, D. N.: Adv. Mass Spectrom. Biochem. Med. *1*, 329 (1976)
111. Smith, A., Pettifor, M. J.: Vacuum *32*, 175 (1982)
112. Spruytenburg, R., Dunn, I. J., Bourne, J. R.: Biotechnol. Bioeng. Symp. *9*, 359 (1979)
113. Srienc, F., Arnold, B., Bailey, J. E.: Biotechnol. Bioeng. *26*, 982 (1984)
114. Stephanopoulos, G, Şan, K. Y.: ibid. *26*, 1176 (1984)
115. Szilágyi, J., Bohátka, S., Langer, G., Sántha, G., Seres, P.: Industrial application of mass spectrometry to monitoring fermentation. In: "Third European Congress on Biotechnology", Vol. 3, p. 609 (DECHEMA ed.), Verlag Chemie, Weinheim 1984
116. Tailliez, B. Y., Hume, S. H.: Dyn. Mass Spectrometry *6*, 181 (1981)
117. Tal'roze, V. L., Gorodetsky, I. G., Zolotny N. B., Karpov, G. V., Skurat, V. E., Maslennikovy, V. Ya.: Adv. Mass Spectrom. *7B*, 858 (1978)
118. Tonge, G. M.: Instrumentation and control in fermentation: The application of computer controlled mass spectrometry. 5th Intern. Ferm. Symp., London, Canada (abstr.) 1980
119. Valentini, L., Razzano, G.: The real-time analysis of broth constituents in the control strategy of the fermentation processes: on/off-line fulfilments. In: "Modelling and control of biotechnological processes", p. 253 (Halme, A., ed.), Pergamon Press, Oxford 1983
120. Van Graas, G., de Leeuw, J. W., Schenck, P. A.: Adv. Organ. Geochem. *12*, 485 (1979)
121. Verduyn, C., Van Dijken, J. P., Scheffers, W. A.: Biotech. Bioeng. *25*, 1049 (1983)
122. Vestal, M. L.: Int. J. Mass Spectrom. Ion Phys. *46*, 193 (1983)
123. Voogd, J., Huitin, E., Van Rossum, G. J., Petri, J. M.: ibid. *48*, 7 (1983)
124. Vorlop, K. D., Becke, J. W., Stock, J., Klein, J.: Anal. Chim. Acta *163*, 287 (1984)
125. Voyksner, R. D.: Anal. Chem. *57*, 2600 (1985)
126. Waller G. R.: Biotechnical applications of Mass spectrometry, Wiley Interscience, New York 1972
127. Wang, H. E., Cooney, C. L., Wang, D. I. C.: Biotechnol. Bioeng. *21*, 975 (1979)
128. Watson, J. T., Biemann, K.: Anal. Chem. *36*, 1135 (1964)
129. Weaver, J. C.: Possible biomedical applications of the volatile enzyme product method. In: "Biomedical Applications of Immobilized Enzymes and Proteins" *2*, p. 207 (Chang, T. M. S., ed.), Plenum Press, New York 1977
130. Weaver, J. C.: Continuous monitoring of volatile metabolites by a mass spectrometer. In: "Non-invasive Probes of Tissue Metabolism", p. 25 (Cohen, J. S., ed.), Wiley, New York 1982
131. Weaver, J. C., Abrams, J. H.: Rev. Sci. Instrum. *50*, 478 (1979)
132. Weaver, J. C., Mason, M. K., Jarrell, J. A., Peterson, J. W.: Biochem. Biophys. Acta *438*, 296 (1976)
133. Weaver, J. C., Perley, C. R. Reames, F. M., Cooney, C. L.: Biotechnol. Letters *2*, 133 (1980)
134. Westover, L. B., Tou, J. C., Mark, J. H.: Anal. Chem. *46*, 568 (1974)
135. Windig, W., de Hoog, G. S., Haverkamp, J.: J. Anal Appl. Pyrolysis *3*, 213 (1982)
136. Windig, W., Kistemaker, P. G., Haverkamp, J.: ibid. *3*, 199 (1982)
137. Woldring, S., Owens, G., Woolford, D.: Science *153*, 885 (1966)
138. Wyatt, J. R.: Int. J. Mass Spectrom. Ion Proc. *60*, 289 (1984)
139. Yamané, T., Matsuda, M., Sada, E.: Biotechnol. Bioeng. *23*, 2493 (1981)

140. Yamané, T., Matsuda, M., Sada, E.: ibid. *23*, 2509 (1981)
141. Yano, T., Kobajashi, T., Shimizu, S.: J. Ferment Technol. *56*, 421 (1978)
142. Zabriskie, D. W., Humphrey, A.: Appl. Environm. Microbiol. *35*, 337 (1978)
143. Villinger, H., Federer, W.: Analysentechnik OHG, Innsbruck, Austria 1986
144. Scott, R. I., Yarlett, N., Hillman, K., Williams, T. N., Williams, A. G., Lloyd, D.: J. Appl. Bacteriol. *55*, 143 (1983)
145. Scott, R. I., Williams, T. N., Lloyd, D.: Biotechnol. Letters *5*, 375 (1983)
146. Lloyd, D., Kristensen, B., Degn, H.: J. Gen. Microbiol. *129*, 2125 (1983)
147. Lloyd, D., Scott, R. I., Williams, T. N.: Trends in Biotechnol. *1*, 60 (1983)
148. Yarlett, N., Scott, R. I., Williams, A. G., Lloyd, D.: J. Appl. Bacteriol. *55*, 359 (1983)
149. Hillman, K., Lloyd, D., Scott, R. I., Williams, A. G.: The effect of O_2 on H_2 production by rumen holotrich protozoa as determined by membrane inlet mass spectrometry. In: "Gas Metabolism", p. 271, (Poole, R. K., Dow, C. S., eds.), Academic Press, New York 1985
150. Scott, R. I., Williams, T. N., Whitmore, T. N., Lloyd, D.: Mass spectrometric determinations of the effect of O_2 on methanogenesis: inhibition or stimulation? In: "Gas Metabolism", p. 263, (Poole, R. K., Dow, C. S., eds.), Academic Press, New York 1985
151. Hillman, K., Lloyd, D., Williams, A. G.: Continuous monitoring of fermentation gases in an artifical rumen system (RUSITEC) using a membrane-inlet probe on a portable quadrupole mass spectrometer. In: "Gas Enzymology", p. 201, (Degn, H., Cox, R. P., Toftlund, H., eds.), Reidel, Dordrecht 1985
152, Whitmore, T. N., Lloyd, D.: Biotechnol. Letters *8*, 203 (1986)
153. Hillman, D., Lloyd, D., Williams, A. G.: Current Microbiol. *12*, 335 (1985)
154. Whitmore, T. N., Lazzari, M., Lloyd, D.: Biotechnol. Letters *7*, 283 (1985)
155. Rannalli, G., Whitmore, T. N., Lloyd, D.: FEMS Microbiol. Letters *35*, 119, (1986)
156. Lloyd, D., Scott, R. I.: Mass spectrometric monitoring of dissolved gases. In: "Gas Metabolism", p. 239 (Poole, R. D., Dow, C. S., eds.), Academic Press, New York 1985
157. Reuss, I., Fröhlich, S., Kramer, B., Messerschmidt, K., Niebelschütz, H.: Mathematical modelling of microbial kinetics and oxygen transfer for gluconic acid production with Aspergillus niger. In: "Third European Congress of Biotechnology" Vol. 2, p. 455 (DECHEMA ed.), Verlag Chemie, Weinheim 1984

Extraction and Purification of Arachidonic Acid Metabolites from Cell Cultures

C. Bedetti and A. Cantafora
Istituto Superiore di Sanità, Viale Regina Elena 299, 00161 Roma/Italy

1 Introduction . . . 48
2 Methods of Extraction . . . 49
2.1 Extraction of Arachidonic Acid Metabolites by Solvents . . . 49
2.2 Extraction of Arachidonic Acid Metabolites by Solid Phases . . . 51
2.2.1 Extraction by Amberlite Resins . . . 51
2.2.2 Extraction of Arachidonic Acid Metabolites by Reverse Phase Cartridge . . . 51
3 Methods of Purification . . . 52
3.1 Separation and Purification of Arachidonic Acid Metabolites by TLC . . . 52
3.2 Separation and Purification of Arachidonic Acid Metabolites by Open-Column Chromatography . . . 57
3.2.1 Straight-Phase Chromatography . . . 58
3.2.2 Reverse-Phase Chromatography . . . 59
3.2.3 Other Stationary Phases for Open-Column Chromatography . . . 59
3.3 Separation and Purification of Arachidonic Acid Metabolites by HPLC . . . 60
3.3.1 Purification of LTs using HPLC . . . 61
3.3.2 Purification of Hydroxy and Hydroperoxy Arachidonic Acid Metabolites Using HPLC . . . 68
3.3.3 Purification of PGs and TXB_2 Using HPLC . . . 69
4 Scale-up Considerations . . . 74
5 Acknowledgement . . . 76
6 List of Abbreviations . . . 76
7 References . . . 77

The oxygenated, biologically active, metabolites of arachidonic acid (5,8,11,14-eicosatetraenoic acid), collectively called eicosanoids, include products formed via the cyclooxygenase enzyme complex (prostaglandins, thromboxanes) and via the lipoxygenases (leukotrienes and hydroxy fatty acids). These compounds can be synthesized in vitro by different types of cells in culture but their analysis for metabolic studies or their isolation for subsequent pharmacological studies require proper procedures of extraction and purification. This paper reviews the various techniques of extraction and emphasizes the major and new procedures. In this connection the use of reverse-phase chromatographic material is particularly stressed both for the extraction of the various classes of metabolites by the solid-phase extraction technique and for the purification of classes or individual compounds by high-performance liquid chromatography.

Advances in Biochemical Engineering/
Biotechnology, Vol. 35
Managing Editor: A. Fiechter

1 Introduction

Arachidonic acid (AA), a polyunsaturated fatty acid, derives in animal metabolism from dietary linoleic acid and is found esterified at the sn-2 position of cellular membrane phospholipids. This compound, when released by phospholipases upon cell stimulation, enters the metabolic pathways of oxygenation collectively known as the AA cascade. The biosynthetic interrelationships among AA and its main oxygenated metabolites called eicosanoids are shown in Fig. 1.

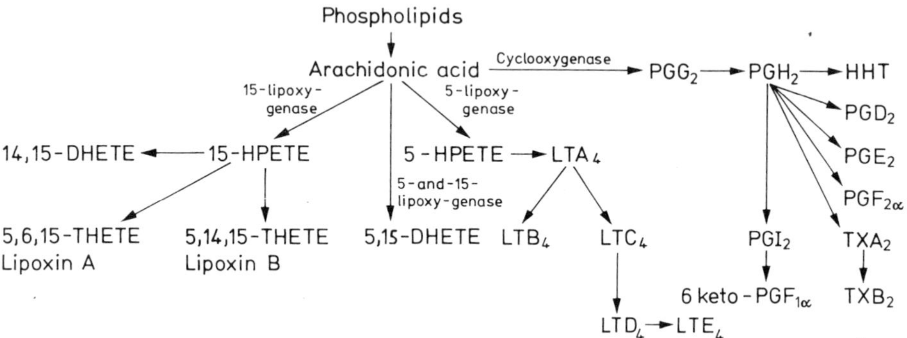

Fig. 1. Metabolic pathways of arachidonic acid in mammalian cells

The growing interest in the different AA metabolites over the past decades reflects a growing recognition of their important roles. Among them we mention the participation of prostaglandins (PGs) in inflammatory reactions, the inhibitory effects of aspirin-like drugs on biosynthesis of PGs, the effects of thromboxane A_2 (TXA_2) and prostacyclin (PGI_2) on platelet aggregation, on platelet–vessel wall interactions and on vessel wall protection, the role of leukotrienes (LTs) in pathophysiological symptoms associated with allergy and inflammation.

The types of AA metabolites vary according to the types of animal species, tissues and experimental conditions but a large number of mammalian cells, in vivo or in vitro, can synthesize some metabolites. The identification and quantification of AA metabolites is essential for an understanding of their biological roles but this task is made difficult by their chemical instability and extremely low concentrations in living systems. This requires very sensitive systems of quantitation, e.g. radioimmunoassay (RIA), gas chromatography/mass spectrometry (GC/MS) or high-performance liquid chromatography (HPLC), and also efficient procedures of extraction and purification for the different biological materials analyzed.

Interest in the methods of extraction and purification is however not restricted to the analytical field. The use of particular cell lines in culture allows the production of large quantities of AA metabolites that are isolated, identified and quantified for subsequent pharmacological and biochemical studies.

This article deals with the major procedures of extraction and purification of AA metabolites from cell cultures described in the literature between 1977 and 1985, with particular consideration of HPLC, which has been increasingly used for both analytical and preparative purposes.

2 Methods of Extraction

2.1 Extraction of Arachidonic Acid Metabolites by Solvents

This type of extraction, although old, is still widely used owing to its simplicity and efficacy.

The AA metabolites are easily extracted from the aqueous environment of cell cultures or tissue homogenates using organic solvents, the recovery rate being improved by acidification of the water phase. The extracts obtained in such a manner are free from salts, proteins and carbohydrates but have to be further purified from other lipids. The methods of extraction by solvent partition can be divided into methods that extract AA metabolites, as well as other lipid classes, without selectivity and methods that extract selectively some specific AA metabolites. The former are very simple but require a further thorough purification. The latter offer extracts sufficiently pure for some analytical applications.

Among the non-specific methods we have to mention the traditional chloroform-methanol (MeOH) extraction procedures of Folch et al. [1] and of Bligh and Dyer [2], which are widely used for extracting all kind of lipids from tissues and body fluids. These procedures have been employed for extraction of the less polar AA metabolites e.g. PGs and TXs. Saunders and Horrocks [3] described the extraction of PGs, TXs and AA from bovine brain with hexane-isopropanol (3:2, v/v) mixtures.

Lipoxygenase products have been extracted from human polymorphonuclear leukocytes (PMN) according to a modification of the method of Bligh and Dyer. Incubations were stopped by acidification with 2 N citric acid to pH 3 and extracted with 3 vol. chloroform-MeOH (2:1, v/v), mixed and separated into two phases by centrifugation at 200 × g for 5 min. The lower chloroform phase was removed and evaporated to dryness under nitrogen [4].

In a simplified method for extraction of cyclooxygenase products from cell cultures, media added with 1 vol. of ethanol (EtOH), brought to pH 3 with formic acid, are extracted twice with 1 vol. each of chloroform. The chloroformic lower phases are combined and taken to dryness under nitrogen [5].

Cyclooxygenase products, LTC, DHETE were extracted from macrophages after acidification with formic acid using chloroform with 0.005% of butylated hydroxy-toluene added as antioxidant [6]. PGD_2, generated by stimulation of murine masto cytoma cell cultures, was extracted with chloroform after acidification of the medium with 1% formic acid [7]. Further purification of PGD_2 for subsequent GC/MS analysis was accomplished by the octadecyl silyl (ODS) cartridge procedure [8]. Suspensions from rabbit PMN were extracted with 8 vol. of chloroform as a preliminary step to LTB_4, 5HETE and 15HETE purification [9].

The extraction with methylene chloride, instead of chloroform, from a media with an appropriate pH value has been used to separate LTC_4, LTD_4 and LTE_4 from DHETEs and 5HETE in samples obtained from incubations with a homogenate of rat basophilic leukemia (RBL-1) cells. Extraction was performed with methylene chloride four times at pH 6.2. The aqueous phase contains LTC_4, LTD_4 and LTE_4; the organic phase contains DHETEs, HETEs and free fatty acids [10]. Actually HETEs and DHETEs contain no peptidic portion and are relatively non-polar. They extract readily from acidified aqueous solutions into the organic phase. The peptide components contribute appreciably to the polarity of LTC_4, LTD_4 and LTE_4, as is reflected by the poor extraction of these compounds in the organic phase.

Other immiscible solvents like diethyl ether or ethyl acetate (EtOAc) have been used to extract selectively some AA metabolites. LTB_4 from human peripheral and rat peritoneal PMN [11–14] and from human leukocytes [15,16] have been extracted with diethyl ether from acidified media. The extraction of AA oxygenation products, including LTB_4, from mouse peritoneal macrophage suspensions has been performed with diethyl ether from media buffered at pH 4.7 [17]. DHETE extraction from rat peritoneal mononuclear cells and human leukocytes has been performed by acidifying to pH 3.5 with 5 N HCl and extracting with 3 volumes of diethyl ether. The extract was washed twice with 0.2 vol. of water, dried over sodium sulfate and taken to dryness [18].

EtOAc has been used [19] for the extraction of PGs from RBL-1 cells: the medium was acidified to pH 3.4 with 2 N formic acid and was extracted 3 times with 1 vol. of EtOAc. A similar procedure was used [20] for the extraction of 15 HETE and DHETE from human eosinophil suspensions: EtOAc and diethyl ether were found equivalent. LTB_4 was extracted for HPLC analysis from alveolar macrophage suspension acidified to pH 4 with 1 N H_3PO_4, saturated with NaCl by 3 partitions with 1 vol. of EtOAc [21].

A different aspect in the use of solvents for the extraction of AA metabolites concerns the use of MeOH or EtOH to disrupt the interactions that bind the highly hydrophilic LTs to the cell structures. In a typical procedure ice-cold EtOH was added to the incubation mixtures for lipid extraction. The mixtures were centrifugated ($1000 \times g$) and the proteinaceus bottom was discarded. The supernatants were evaporated to dryness under nitrogen [22]. For a better LTB_4 extraction, the samples (medium-EtOH mixtures 1:1, v/v) were spun down to remove precipitated protein, then acidified to pH 3 with 85% formic acid and extracted twice by 1 vol. chloroform. The aqueous phase was brought to neutral pH by NH_4OH and the remaining EtOH was evaporated under nitrogen. The samples were then adjusted to pH 4 and extracted twice with EtOAc. The organic phases were pooled together, and after addition of NH_4OH were dried under nitrogen. The extraction efficiency was over 90% for LTB_4 [23]. According to the same authors the cells were extracted for determination of intracellular lipoxygenase products by 4 ml of chloroform-MeOH dried and digested overnight with MeOH—NaOH 0.1 N (1:1, v/v). MeOH was dried and the aqueous phase was extracted twice with EtOAc after adjusting the pH to 4.

The following procedure is an example of selective extraction of LTB_4 from human leukocyte incubation mixtures. The acidified incubation mixture was partitioned twice with EtOAc and following centrifugation the two phases were evaporated. The residue was resuspended in 0.1 M sodium phosphate buffer, pH 8.4, and partitioned twice with 1-chlorobutane. In this system monohydroxy fatty acids and less polar material are preferentially soluble in the organic phase; whereas LTB_4 and more polar material remain in the aqueous phase. The aqueous phase was then acidified with glacial acetic acid (AcCOOH) and extracted twice with EtOAc. The pooled EtOAc was evaporated and stored at −20 °C in MeOH [24].

Extractions are generally carried out in siliconized glassware to minimize non-specific binding of lipids (particularly LTs).

2.2 Extraction of Arachidonic Acid Metabolites by Solid Phases

2.2.1 Extraction by Amberlite Resins

Extraction by Amberlite resins is more selective than solvent extraction; it has been widely used in recent years for the recovery of PGs from body fluids. Amberlite resins absorb selectively PGs or TXs from aqueous solutions acidified to pH 3.0–3.5, because organic compounds when ionized bind less to Amberlite resins, and thus the binding of AA metabolites may be increased by lowering the pH of the aqueous solution. Amberlite resins are then eluted with MeOH.

These resins require preparation prior to their use: they have to be soaked with 95% EtOH and then they have to be washed many times with water in order to remove the EtOH completely. This material is packed in a short chromatography column (1.5 × 15 cm length) and stored with water.

The Amberlite resin XAD-2 was introduced first. Its use has been described [25] for the extraction of PGs from acidified plasma. It has been most frequently employed in extraction of PGs from body fluids that contain low concentrations of compounds in large volumes in order to reduce the amount of toxic and inflammable solvents required by other techniques.

XAD-7 and XAD-8 have been used for extraction of sulfidopeptide LTs from cell cultures including murine mastocytoma cells [26], RBL-1 cells [27–29], and eosinophils [30]. It was observed that the recovery rate is affected by impurities present in some batches and that the resin requires extensive preparation, which is time-consuming in comparison to the most recent reverse-phase (RP) extraction procedures [31].

Moreover, it was observed that significant amounts of polymer may elute from the Amberlite column. This severely affects the binding of PGs to specific antibodies [32] and for this reason it is not advisable to use them in preparing samples for RIA.

2.2.2 Extraction of Arachidonic Acid Metabolites by Reverse-Phase Cartridge

This recent technique, based on the same principle as RP chromatography, has found favour with researchers over the past few years. The RP cartridges, generally made with ODS phases, are columns with low efficiency and high selectivity. They remove efficiently from water solutions a wide range of substances containing a hydrophobic side, e.g. the AA metabolites, and by elution with a strong organic solvent, e.g. MeOH, the extracted substances are transferred to a small volume of a volatile solvent. The extracts prepared in this manner generally contain high levels of contaminants but by introducing intermediate washings before the elution with pure MeOH it is possible to have purified extracts and fractions based on large differences in polarity.

For the simple extraction of all AA metabolites from aqueous solutions the major advantages over traditional solvent extractions are:

1) only small amounts of solvents are required;
2) the extracts are free from water and are easily taken to dryness under nitrogen.

Many applications to cell suspensions or to incubation media have been described: e.g. LTB_4 and other DHETEs produced by cell-free enzyme systems of RBL-1 cells have been extracted by passing the sample through an ODS cartridge and eluting it with MeOH after washing with pure water [33]. Similar procedures using a C18 RP extraction column (J. T. Baker, Deventer, The Netherlands) have been described [34]. LTC_4, LTD_4 and LTE_4 standards were quantitatively recovered from a saline solution [35] by passing it through a sep-pak cartridge and eluting with 90% MeOH in water.

LTs generated by human eosinophils in response to IgG-dependent stimulation have been purified [36] by applying extracts in 10% MeOH, acidified to pH 4, to a sep-pak ODS cartridge and eluting the cartridge with 100% MeOH.

More frequently used is the procedure of extraction and purification by stepped elution of fractions with different polarity. One of the first applications of this type has been described [8] for the purification of PGs and related compounds from body fluids and tissues: the supernatant of an aqueous homogenate, acidified to pH 3 with HCl, is passed through a sep-pak ODS cartridge that is eluted successively with 20 ml EtOH-water (15:85, v/v), 20 ml petroleum ether and 10 ml methyl formate. PGs and TXs are recovered in high yield (84%–100%) and purity in this last fraction.

Extracts from RBL-1 cells have been applied to a C18 sep-pak cartridge previously treated with EtOH and water. After the sample application the cartridge was washed with water and 15% EtOH (v/v). LTs and HETEs were eluted with methyl formate [37].

Conversion of LTA_4 by human erythrocytes was analyzed by purifying the supernatant on a C18 sep-pak cartridge and subsequently eluted with water, hexane, EtOAc and MeOH. The fractions eluted with EtOAc and MeOH contained respectively LTB_4 and LTC_4 [38]. LTB_4 and its ω-oxidation products were extracted and purified for subsequent HPLC analysis from human PMN incubation medium: the sample was acidified to pH 3, applied to the cartridge and eluted sequentially with 10% EtOH (v/v), chloroform–petroleum ether (35:65, v/v) mixture and methyl formate; this last fraction contained LTB_4, 20-COOH—LTB_4 and 20-OH—LTB_4 but was relatively free of polar phospholipids, AA and HETEs. The recoveries of 20-COOH—LTB_4, 20-OH—LTB_4, PGB_2 and LTB_4 were respectively of the order of 81%, 84%, 79% and 82% [39].

Extracts from RBL-1 cells were dissolved in MeOH-water (20:80, v/v), acidified by the addition of 1% AcCOOH, and purified by passing it through a J. T. Baker ODS column and eluting subsequently with MeOH-water (50:50, v/v), MeOH-water (75:25, v/v) and pure MeOH. The last two fractions contained respectively PGs and LTs with related compounds. The recoveries determined with standard compounds were in the order of 95%–99% for PGE_2, PGD_2, LTD_4, 15 HETE and 15 HPETE [40].

3 Methods of Purification

3.1 Separation and Purification of Arachidonic Acid Metabolites by TLC

Thin layer chromatography (TLC) has been used right from the beginning in research into prostaglandins [41, 42] and is still used for both analytical and preparative applications because of its simplicity and inexpensiveness. The major groups of prostaglandins (A, B, D, E, F, 6-keto-$PGF_{1\alpha}$) as well as TXB_2 and hydroxyderivatives of AA are readily separated on silica gel G using various solvent systems, as shown in Table 1.

Table 1. Separation of arachidonic acid metabolites using TLC

Arachidonic acid metabolites	Source	Species	Solvent system	Ref.
PGE$_2$, PGF$_{2a}$, TXB$_2$	Platelets	Human	Hexane/ether/AcCOOH (70:30:1) EtOAc/MeOH/water (80:20:50)	[43]
TXB$_2$, HETE, HHT	Platelets	Human	EtOAc/isooctane/water (50:100:100)	[44]
PGs, TXB$_2$, HETE	Platelets	Human	Chloroform/MeOH/AcCOOH (180:10:10) Heptane/ether/AcCOOH (60:40:1)	[45, 46]
TXB$_2$, HETE, HHT	Platelets	Human	Hexane/ether/AcCOOH (50:50:1)	[47]
PGs, PGI$_2$ (sodium salt)	Platelets	Human	EtOAc/isooctane/AcCOOH/water (110:50:20:100)	[48]
TXB$_2$, HETE	Platelets	Human	Hexane/ether/AcCOOH (60:40:1) Ether/MeOH/AcCOOH (90:1:2)	[49]
TXB$_2$	Platelets	Human	EtOAc/isooctane/AcCOOH/water (110:50:20:100)	[50]
TXB$_2$, PGF$_{2a}$, PGE$_2$, HETE	Platelets	Human	Chloroform/MeOH/AcCOOH/water (90:8:1:0.8)	[51, 52]
TXB$_2$, HHT	Platelets	Human	EtOAc/isooctane/AcCOOH/water (90:50:20:100)	[53]
TXB$_2$	Platelets	Human	Chloroform/EtOAc/MeOH/AcCOOH/water (70:30:8:1:0.5)	[54]
TXB$_2$, HHT, HETE	Platelets	Human	Chloroform/AcCOOH (90:3) Ether/MeOH/AcCOOH (90:1:2) EtOAc/AcCOOH (99:1)	[55]
TXB$_2$, HETE, THETE	Platelets	Rat	EtOAc/isooctane/AcCOOH/water (110:50:20:100)	[56, 57]
TXB$_2$, PGE$_2$, PGF$_{2a}$	Platelets	Rabbit	Chloroform/MeOH/AcCOOH/water (90:8:1:0.8)	[58]
LTB$_4$, 5 HETE, DHETE	Peripheral blood PMN	Human	Chloroform/MeOH/AcCOOH/water (90:8:1:0.8)	[59]
LTB$_4$, 5 HETE	Peripheral blood PMN	Human	Ether/petroleum ether/AcCOOH (50:50:1) (LTB$_4$, 5HETE)	[60]
PGs, TXB$_2$			EtOAc/isooctane/AcCOOH/water (110:50:20:100) (PGs, TXB$_2$) EtOAc/AcCOOH (99:1)	
LTB$_4$, 20-OH—LTB$_4$	Peripheral blood PMN	Human	EtOAc/isooctane/AcCOOH/water (110:50:20:100)	[61]
HETEs, HPETEs	Peripheral blood PMN	Human	Ether/petroleum ether/AcCOOH (50:50:1)	[4, 62]
5 HETE, LTB$_4$	Peripheral blood PMN	Human	Ether/petroleum ether/AcCOOH (60:40:1)	[63]
5 HETE, 5,12 DHETE	Peritoneal PMN	Rabbit	Ether/petroleum ether/AcCOOH (50:50:1)	[9]
DHETEs	Peritoneal PMN	Rabbit	Ether/hexane (70:30)	[64]
5 HETE, LTB$_4$	Peritoneal PMN	Rat	Ether/ligroine/AcCOOH (60:40:1)	[65]
TXB$_2$, 5 HETE, HHT	Peritoneal PMN	Rat	EtOAc/isooctane/AcCOOH/water (110:50:20:100)	[66]

Table 1. (continued)

Arachidonic acid metabolites	Source	Species	Solvent system	Ref.
12 HETE, PGs, TXB_2	Peripheral blood monocytes	Human	EtOAc/isooctane/AcCOOH/water (90:50:20:100) Benzene/dioxane/AcCOOH (200:100:1)	67)
12 HETE	Peripheral blood leukocytes	Pig	Ether/petroleum ether/AcCOOH (85:15:0.1)	68
Lipoxins	Peripheral blood leukocytes	Human	EtOAc/isooctane (5:1)	69)
TXB_2, PGs, 6-Keto-PGF_{1a}	Peritoneal cells	Rat	EtOAc/AcCOOH (99:1) EtOAc/isooctane/AcCOOH/water (110:50:20:100)	56)
PGs, 6-Keto-PGF_{1a}	Peritoneal cells	Rat	EtOAc/isooctane/AcCOOH/water (110:50:20:100) Chloroform/MeOH/AcCOOH/water (90:8:1:0.75)	70)
LTC_4, PGE_2	Peritoneal cells	Mouse	EtOAc/MeOH/AcCOOH (95:5:1) Ether/hexane/AcCOOH (40:60:1)	17)
PGs, 6-keto-$PGF_{1\alpha}$	Peritoneal cells	Mouse	EtOAc/isooctane/AcCOOH/water (110:50:20:100)	71, 72)
PGs, 12 HETE	Alveolar macrophages	Rabbit	EtOAc/isooctane/AcCOOH/water (110:50:20:100)	73)
LTB_4, HETE, PGs, TXB_2	Alveolar macrophages	Rat	Ether/petroleum ether/AcCOOH (50:50:1) EtOAc/isooctane/AcCOOH/water (110:50:20:100) EtOAc/AcCOOH (99:1)	74)
PGs, 6-Keto-PGF_{1a}	Smooth muscle cells	Human	Chloroform/MeOH/AcCOOH/water (90:8:1:0.8) EtOAc/isooctane/AcCOOH/water (110:50:20:100)	75)
PGE_2, PGF_{2a}, 6-Keto-PGF_{1a}	Smooth muscle cells	Rabbit	EtOAc/isooctane/AcCOOH/water (90:50:20:100) EtOAc/isooctane/AcCOOH/water (110:50:20:100)	76, 77)
HETEs	Smooth muscle cells	Rat	Ether/ligroine/AcCOOH (50:50:1)	78)
6-Keto-PGF_{1a}, PGs, HETE	Endothelial cells	Human	EtOAc/isooctane/AcCOOH/water (110:50:20:100)	79–81)
6-Keto-PGF_{1a}, PGE_2	Endothelial cells	Bovine	EtOAc/isooctane/AcCOOH/water (110:50:20:100)	82)
PGs, 6-Keto-PGF_{1a}	Mesothelial cells	Rabbit	EtOAc/isooctane/AcCOOH/water (110:50:20:100)	83)
PGs, 6-Keto-PGF_{1a}, TXB_2, HETE	Lung cells	Human	Toluene/dioxane/AcCOOH (65:34:1.5)	84)
5 HETE	Bone marrow/mast cells	Mouse	Ether/petroleum ether/AcCOOH (50:50:1)	85)
5 HETE	PT18 cells	Mouse	Ether/petroleum ether/AcCOOH (50:50:1)	86, 87)
5 HETE, DHETE, PGD_2, TXB_2	MC9 mast cells	Mouse	Ether/ligroine/AcCOOH/water (60:40:1) EtOAc/isooctane/AcCOOH/water (90:50:20:100) Chloroform/MeOH/AcCOOH/water (90:8:1:0.8)	22)
PGs	Macrophage cell lines	Mouse	EtOAc/isooctane/AcCOOH/water (110:50:20:100)	88)
HETE, DHETE, PGs	RBL-1 cells	Rat	EtOAc/isooctane/AcCOOH/water (110:50:20:100)	89–91)

PGs	RBL-1 cells	Rat	Chloroform/MeOH/AcCOOH/water (90:8:1:0.8)	[19]
			EtOAc/isooctane/AcCOOH/water (90:50:20:100)	
			Benzene/dioxane/AcCOOH (60:30:3)	
PGD_2, 5 HETE, DHETE	RBL-1 cells	Rat	EtOAc/isooctane/water (50:100:100)	[92]
			Benzene/ether/EtOH/AcCOOH (50:40:2:0.2)	
5 HPETE	RBL-1 cells	Rat	Ether/petroleum ether/AcCOOH (85:15:0.1)	[93]
HETE, PGs	RBL-1 cells	Rat	Ether/petroleum ether/AcCOOH (55:45:1)	[94]
			Benzene/dioxane/AcCOOH (20:10:1)	

Precoated TLC plates are now commercially available from many sources, but in the past most investigators used to prepare their own thin layer plates. The commercial plates available are coated on glass, aluminium foil, or plastic; plates with a fluorescent indicator were sometimes used to make UV absorbing compounds visible.

For the recovery of substances from commercial plates it is important first to wash the plate with chloroform-MeOH (2:1, v/v). It is advisable to activate the plate at 110 °C for 30 min before use. TLC plates (5 × 20 cm) are developed generally in cylindrical tanks containing 30 ml of solvent. Large plates (20 × 20 cm) are developed in rectangular tanks containing 200 ml of solvent.

Single-step chromatography of AA metabolites has traditionally been accomplished using the organic phase of EtOAc-isooctane-AcCOOH-water (110:50:20:100, v/v) mixture [42)] as the solvent system for extracts from platelets [48, 50, 56, 57)], PMN [61, 66)], peritoneal macrophages [71, 72)], alveolar macrophages [73)], vascular endothelial cells [79–82)], tumor-derived murine macrophage cell lines [88)] and the RBL-1 cell line [89–91)]. Slightly different solvent systems have also been used [53, 67)].

The separation of PGA_1, PGB_1, PGB_2, PGD_2, $PGE_{1\alpha}PGE_2$, $PGF_{1\alpha}$, $PGF_{2\alpha}$ and 6-keto-$PGF_{1\alpha}$ was compared using seven solvent systems [95)]. This mixture spotted on TLC silica gel plates (10 µg each) was resolved into its components when developed in chloroform-MeOH—AcCOOH (90:5:5, v/v).

Extracts from human smooth muscle and endothelial cells [75)], human platelets and neutrophils [51, 52, 59)] and rabbit platelets [58)] have been separated chromatographically with chloroform-MeOH—AcCOOH-water (90:8:1:0.8, v/v). Some investigators used a double developing system to increase resolution of components of extracts.

A combination of solvent systems is necessary to confirm the presence of a particular PG since many metabolites of PGs may have polarities comparable with that of the parent compound. Actually PGE_2 and TXB_2 are not clearly resolved with the solvent system EtOAc-isooctane-AcCOOH-water as well, as $PGF_{2\alpha}$ may partially comigrate with 13,14-dihydro-6-keto-$PGF_{1\alpha}$. A better separation of these last compounds was obtained [82)] with silica gel plates impregnated with boric acid. Several systems, including chloroform-MeOH—AcCOOH, do not permit separation of PGE from 6-keto-PGF_1, and LTB_4 from PGD_2 and 12 HETE from 12 HPETE [59)]. The hydroperoxy- and hydroxyeicosatetraenoic acids, including LTB_4 are better separated by TLC using a combination of diethyl ether–hexane–AcCOOH [47, 49)].

Extracts have also been purified with a solvent system of diethyl ether–petroleum ether–AcCOOH, in different ratios [4, 9, 60, 62, 63, 68, 85–87, 93)], which resolves positional isomers of hydroxy derivatives of AA.

To protect hydroperoxy-derivatives of AA from spontaneous decomposition, the acidified reaction mixture was extracted with ice cold diethyl ether and subjected to TLC at −10 °C in a solvent system of diethyl ether–petroleum ether–AcCOOH [93)].

The complete separation of each of the individual TXs, PGs and hydroxy fatty acids on a single TLC plate has been achieved by two-dimensional TLC procedures. For instance, the plate has been developed in the first direction with chloroform-AcCOOH (90:3, v/v) which moved the hydroxy fatty acids and arachidonic acid well up the plate and left the TX and PGs close to the origin, and in the second direction with diethyl ether–MeOH–AcCOOH (90:1:2, v/v), which separated TXB_2 from the PGs [55)].

Polar compounds and ω-oxidized metabolites have been separated by a combination of EtOAc—AcCOOH-isooctane and chloroform-MeOH—AcCOOH. This solvent combination also gives a good separation between PGE_2 and TXB_2.

Detection of arachidonic acid metabolites fractionated on the plates may be performed using either non-destructive or destructive reagents. With regard to the first type, iodine vapour staining produces yellow-brown spots and this should reveal microgram amounts of arachidonic acid metabolites [5, 9, 19, 45, 47, 48, 59, 61, 67, 70, 72, 75–77, 82, 88–90, 94]. Phosphomolybdic acid 10% in EtOH has also been used by many researchers [43, 55, 57, 70, 81, 83, 96] to make the standards visible.

TXB_2 and PGs appear as blue spots on a yellow background when plates are sprayed with phosphomolybdic acid in EtOH and heated for about 5 min at 100 °C. Color reagents were considered useful for identification of PGs in TLC plates since the combination of chromatographic mobility and specific color reactions can make the identification of PGs on TLC easier. When TLC plates were sprayed with anisaldehyde–sulfuric acid reagent different colored spots were given by PGs: orange-red (PGA_1, PGB_2), orange-brown (PGD_2), brown (PGE_1, PGE_2), blue-violet ($PGF_{1\alpha}$, $PGF_{2\alpha}$) and yellow (6-keto-$PGF_{1\alpha}$). The limit of detection was found to be 0.1 µg or more.

On plates sprayed with the acidic reagent 2,4-dinitrophenylhydrazine, the carboxylated PGE_1, PGE_2 and PGD_2 progressively developed a yellow-orange chromogen at room temperature, the color development being accelerated by heating. PGD_2 was more reactive than PGE. Also, colored compounds were formed with major PGs after spraying TLC silica gel plates with cupric acetate–phosphoric acid reagent and heating at 85 °C [95].

A solution of 8-hydroxy-1,3,6 pyrene trisulfonic acid trisodium salt (10 mg in 100 ml MeOH) was found [97] to be sensitive for locating PGs (as low as 200 ng) on TLC and at the same time it is a non-destructive spray.

The concentration of AA metabolites in cell extracts is often below the limit of detection on spraying the plate with some TLC reagent. In these cases radiolabelled compounds are used in connection with techniques such as autoradiography, radiochromatogram scanning or liquid scintillation counting of scraped zones.

Elution of fractionated compounds from TLC plates requires care to obtain good recoveries. PGs are usually eluted with MeOH or EtOH, and HETEs with diethyl ether. High recoveries of AA metabolites were obtained when chromatography was carried out on them as methyl esters. Thus AA was extracted from the cell and esterified prior to TLC. The esters of the various classes of AA metabolites were separated in solvent systems similar to those described for the free acids except that the solvent system was not acidified [64].

3.2 Separation and Purification of Arachidonic Acid Metabolites by Open-Column Chromatography

Open-column chromatography is an "old" technique used for separating and purifying AA metabolites. It was applied in the 1960s for isolation and identification of PGs from human seminal plasma [98]. It has been based from the beginning on either straight-phase (SP) or reverse-phase (RP) partition mechanisms using respectively non-treated and silanized silica gels as stationary phases. More recently other types of stationary phases have been introduced including ion-exchange and gel-filtration resins. In addition, the field of application of this technique has been extended from PGs and TXs to HETEs, DHETEs and LTs. The introduction of HPLC has reduced interest in open-column chromatography, but it remains a useful tool for cleaning up complex mixtures and a low-cost alternative to the most recent techniques, when it is employed in sequential combination with a XAD resin column.

3.2.1 Straight-Phase Chromatography

Straight-phase (SP) chromatography with silicic acid columns is conveniently used to separate PGs. By means of silicic acid 80–100 mesh, previously activated at 120 °C for 1 h and mobile phases made with EtOAc-benzene mixtures, PGs and related compounds can be fractionated. The mixture EtOAc-benzene in the ratio 3:7 (v/v) elutes PGA_2 and PGB_2 compounds; on increasing the ratio to 6:4 (v/v) PGE_2 compounds are eluted, and with a further increase to the ratio 8:2 (v/v) $PGF_{2\alpha}$, 19-hydroxy PGA_2 and 19-hydroxy PGB_2 are recovered in eluates [42, 98]. Benzene has been replaced by toluene [95]; by using increasing amounts of EtOAc in toluene (40%, 60% and 80%), PGB_2, PGE_2 and $PGF_{2\alpha}$ eluted in that order. The PGI_2 metabolite 6-keto-$PGF_{1\alpha}$ that remained firmly bound to the silica gel was effectively removed by pure MeOH. Another example is the separation of the major metabolites released from platelets incubated with AA [44]: unreacted AA was eluted with diethyl ether–light petroleum (1:9, v/v), HETE and HHT were eluted, respectively, with ratios of 25:75 (v/v) and 40:60 (v/v) of diethyl ether–light petroleum and then TXB_2 with EtOAc.

Silicic acid column chromatography has also been used to isolate sulfidopeptide LTs from other AA metabolites; a fraction containing AA, HETEs, DHETEs and PGs was eluted with EtOAc—MeOH (10:1, v/v) and sulfidopeptide LTs with MeOH–0.3% aqueous EDTA (95:5, v/v) [22].

The group separation of cysteinyl LTs and hydroxy acids lacking the peptide substituent has been achieved by means of chromatography on a silicic acid column. A certain overlap of trihydroxy acid and dicarboxylic acids and the cysteinyl-containing LTs may occur because metabolites formed through the 5-lipoxygenase pathway can build δ-lactones between the 5-hydroxyl group and the carboxyl group of the AA moiety, and thus they behave as if they have one carboxylic group less [99].

This kind of separation was performed [26] by a series of elutions with diethyl ether–hexane (3:7, v/v) EtOAc, 5% MeOH, 10% MeOH, 50% MeOH EtOAc (v/v) and 100% MeOH. The EtOAc fraction contains LTB_4 and DHETE.

LTD_4 was purified from RBL-1 cell extracts by using a silicic acid column prepared with chloroform and successively eluted with chloroform, chloroform-MeOH (95:5, v/v), MeOH and MeOH-water (99:1, v/v). LTs were recovered in the last two fractions [28]. Purification of the same type of cell extracts was also achieved by means of elution of LTD_4 and LTE_4 with MeOH—EtOAc (1:1, v/v) mixture and of LTC_4 with pure MeOH [100].

Systems based on diethyl ether–hexane mixtures followed by EtOAc have been used for isolation of HETEs, DHETEs and LTB_4 by many authors [11, 101, 102]. HETEs and DHETEs were fractionated from human PMN after ether extraction using a column (0.5 cm i.d.) packed with 1 g of silicic acid: the unreacted AA was eluted with diethyl ether–hexane (20:80, v/v) and HETE and DHETE were eluted with 30 ml of pure EtOAc [101]. Following a similar procedure, LTB_4 was recovered from human PMN extracts in the EtOAc fraction [11].

The ω-oxidized metabolites of LTB_4 were fractioned [102] as described [16] for LTB_4, DHETEs and trihydroxy-eicosatetraenoic acid (THETE), using the following mixtures: diethyl ether–hexane (10:90, v/v) and (40:60, v/v) followed by MeOH—EtOAc (5:95, v/v). Another example is the separation of hydroxy-derivatives of AA formed after addition of AA to suspensions of human blood leukocytes and platelets. HETE and DHETE were isolated from unreacted AA and other polar lipids by eluting the latter with diethyl ether–hexane (10:90, v/v) and the former with diethyl ether–MeOH (95:5, v/v) [103].

3.2.2 Reverse-Phase Chromatography

Reverse-phase open-column chromatography has made a valid contribution to the study of AA metabolites in recent years. The introduction of efficient chemically bound phases has increased interest in chromatographic techniques utilizing the reversed-phase partition principle such as RP-HPLC (see below) and RP cartridges (Sect. 2.2.2). A classical technique of RP open-column chromatography, used at the Karolinska Institute in Stockholm, has been described [25]. The stationary phase is prepared by silanizing the "Hyflo-Supercel" celite that is slurry packed with iso-octanol-chloroform mixtures. Many PGs and their metabolites are fractionated with eluents of MeOH-water mixtures with 38%–50% of MeOH.

3.2.3 Other Stationary Phases for Open-Column Chromatography

Ion-exchange open-column chromatography has been used for the purification (by a DEAE cellulose column) of LTs released from RBL-1 cells. This technique, which is useful for preparing samples for HLPC analysis, fractionates LTs from other AA metabolites by eluting successively with chloroform-MeOH (7:3, v/v), chloroform-MeOH (3:7, v/v), MeOH and MeOH containing 10% (v/v) of 0.5, 0.2, 1.0 and 2.0 M $(NH_4)_2CO_3$ in water [104].

Another example of a purification procedure using an ionic exchange lipophilic gel has been described for the purification of LTs and lipoxygenase products from rat mononuclear cells. Fractions showing slow reacting substance (SRS) biological activity in MeOH-water (72:28, v/v) were applied to a column of DEAE-LH20 equilibrated with the same solvent and eluted as described [99]. The fractions containing the LTs and lipoxygenase products were further applied to a lipophilic gel Sephadex LH-20 column and eluted with chloroform-MeOH (1:1, v/v) mixture [105].

Sequential absorption, gel filtration and partition chromatography on Sephadex LH-20 was used for LT purification from RBL-1 cells and rat peritoneal mast cells. LTs absorb onto Sephadex LH-20. By passing solvents of increasing polarity through a LH-20 column a good separation of LTs from nonpolar lipids is achieved. Butanol-water (14:1, v/v) was effective in releasing both mast cell and RBL-1 cell LTs from Sephadex LH-20 absorbent. Partition chromatography resulted in consistent losses of RBL-1 and mast cell SRS activity. These losses of activity may be caused by the effective pH of the butanol-water solvent. This was 6.5–7.0 during partition chromatography and 3.0 to 8.5 during absorption chromatography. Furthermore, the time required for absorption chromatography was shorter than that required for partition chromatography [106].

3.3 Separation and Purification of Arachidonic Acid Metabolites by HPLC

HPLC has been extensively used for the separation and purification of AA metabolites from various biological sources. This method has the disadvantage that it requires expensive equipment, but, on the other hand, it allows the separation and the purification of closely related compounds with a degree of efficiency, selectivity and reproducibility not attainable by means of other techniques. As far as sulfidopeptide LTs are concerned there are no real alternatives to this technique, while for PGs, TXs and HETEs open-column chromatography or TLC can offer a less efficient but more economical alternative.

Many types of stationary phases for column packing are commercially available but only two of them have been widely used: silica gel for the SP-HPLC stationary phase and the chemically bonded ODS or C18 stationary phase for RP-HPLC.

The AA metabolites having in their molecules both lipophilic chains and hydrophilic functional groups (carboxyl, carbonyl, hydroxyl) theoretically allow separation by both SP-HPLC and RP-HPLC. The combination of SP and RP-HPLC in sequence provides an excellent method for achieving purification to the high degree required for MS analysis. RP-HPLC has been demonstrated to be more versatile and practical and so has gained in favour. By gradient elution RP-HPLC in fact a wide spectrum of individual molecular compounds in samples containing all the classes of AA metabolites can be fractionated. Their detection during chromatography can be achieved by continuous monitoring of the UV absorption. Registration during chromatography of the complete UV spectrum of a chromatographic peak helps to establish its identity and purity. This can be done either using the stopped flow technique with scanning UV spectrophotometer or using the continuous flow technique with a computerized diode array spectrophotometer as detector in the chromatography.

The structural identity of lipoxygenase products was often confirmed by their retention times on SP and RP-HPLC in comparison with authentic standards.

It must be pointed out that notwithstanding the high degree of resolution of HPLC columns, samples require a clean-up step prior to injection. The levels of AA metabolites are in fact low in crude extracts and the cleanup avoids the loss of resolution caused by overloading the column with extraneous material. The cleanup reduces the level of interference that can make the UV detectors blind in the range of short wavelengths (190–210 nm). Moreover, the clean-up extends the life of the column by avoiding rapid deterioration of the head due to irreversible binding of extraneous material. Therefore the use of a clean-up step as described in the sections on solid-phase extraction and open-column chromatography is advisable. Besides, it is also advisable to filter the sample with filtering membranes to avoid clogging of septa with particulate material and to take care of the column (overnight solvents and washing cycles) as advised by the manufacturers.

Each class of compounds however has peculiar problems of fractionation, detection and purification that will be discussed separately in the following sections.

3.3.1 Purification of LTs using HPLC

RP-HPLC was fundamental in determining the isolation and identification of LTs [26, 107, 108]. In fact, by means of this procedure the LTs are easily separated from other AA metabolites and characterized on the basis of their UV spectra when using the scanning UV detector. Their UV detection is more practical than that of PGs and TXs. This is due to the relatively high extinction coefficient of the triene chromophore around 280 nm, allowing UV detection at 280 nm which minimizes interference from compounds which absorb in the low UV but may coelute with LTs (Table 2). Thus the use of an internal standard in HPLC analysis enabled quantitation of these products. The main advantage of HPLC analysis as an analytical method is the possibility of profiling AA metabolites; its main drawback is the relatively high limit of detection (1–5 ng of compound using UV photometry) (Fig. 2).

RP-HPLC has been used for purification and fractionation of cysteinyl containing LTs from various sources including principally RBL-1 cells [27–29, 33, 100, 112–116], alveolar macrophages [117–121], mast cells differentiated in vitro from bone marrow [122, 123], eosinophils [30, 36], leukocytes [34], peritoneal mouse macrophages [124, 125] and murine mastocytoma cells [7, 22, 26, 126–128].

The chromatographic behavior of sulfidopeptide LTs is greatly affected by the polarity and pH of the solvent systems used, due to the presence of acidic carboxyl groups in their molecules. A lesser influence has to be attributed also to differences in the commercial origins of the RP columns. Moreover the elution time required for separation with a radial compression column has been shown to be shorter than that with a conventionally packed column [121].

It is noteworthy however that many authors used very similar chromatographic conditions [7, 26, 28, 30, 115, 117–119, 122, 123, 126, 128]: RP type C18 with 5 μm diameter particles and solvent system MeOH-water (65:35, v/v) acidified with AcCOOH to pH = 5.5. The elution order in such conditions is LTC_4, D_4 and E_4; 11-*trans* isomers elute immediately after 11-*cis* analogues. Minor modifications to this system [27, 33, 36, 70, 100, 113] have been suggested, probably due to the differences in the commercially available C18 stationary phase.

Addition of an excess of AcCOOH and NH_4OH (beyond that required to achieve an optimal pH) was found [31] to maximize peak sharpness and to meliorate selectivity and capacity. This effect may be due to the increase of ionic strength of the mobile phase.

The concentration and type of salts can affect the resolution of LTC, D, E on account of silanol residues of an ODS column interfering with the peptide or amino acid portion

Table 2. Ultraviolet absorption data. Molar extinction coefficients (ε) and maximum wavelength λ_{max} (nm) in MeOH

Compounds	ε	λ_{max} (nm)	Ref.
LTA_4 (methyl ester)	40.000	278	109)
LTB_4	50.000	270	110)
Cysteinyl containing LTs	40.000	278–280	111)
HETEs	30.000	235	101)
DHETE	39.000	281	101)

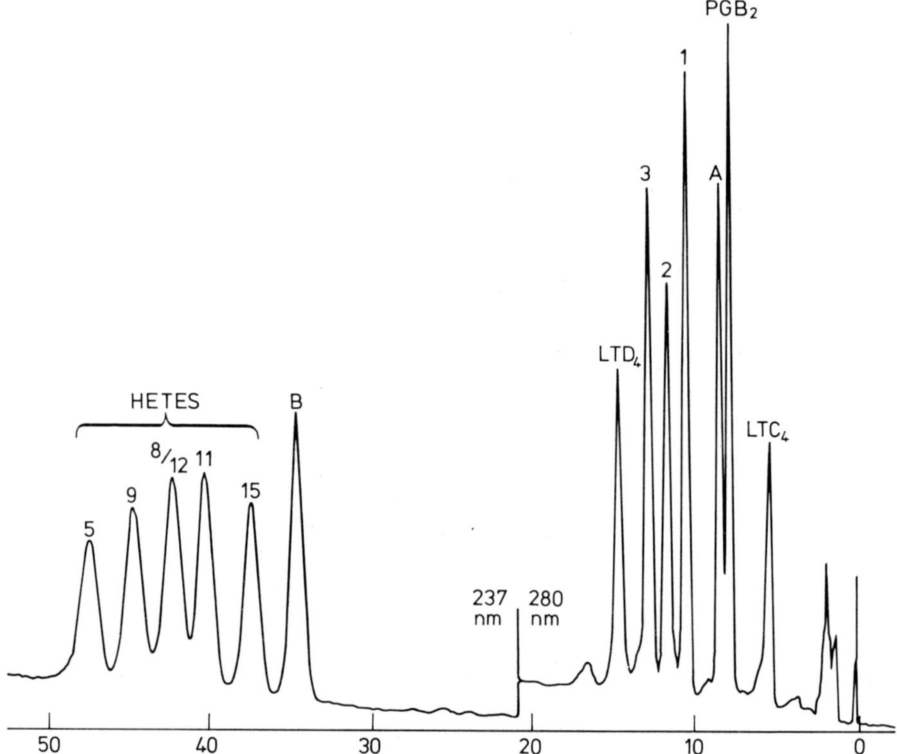

Fig. 2. HPLC Chromatogram of lipoxygenase products (purified standards). Separation achieved on Techsphere 5 C18 (12.5 cm × 4.9 mm i.d.) eluted with methanol water acetic acid (65:35:0.06), pH* adjusted to 5.3 with ammonia [125].
A — 4S,11R + 4R,11S-dihydroxy-5E,7E,9E,13Z-nonadecatetraenoic acid
B — 13 hydroxy-linoleic acid
1 — 5S,12R,dihydroxy-6E,8E,10E,14Z-eicosatetraenoic acid
2 — 5S,12S,dihydroxy-6E,8E,10E,14Z-eicosatetraenoic acid
3 — 5S,12R,dihydroxy-6Z,8E,10E,14Z-eicosatetraenoic acid

of LTs [129]. When a mobile phase of acetonitrile (AcCN)-water-MeOH with 1% AcCOOH and adjusted to pH 5.6 with NH_4OH or triethylamine was used, excellent resolution of LTs was obtained when compared with the results obtained using a mobile phase with 0.1% AcCOOH. The increment of AcCOOH—NH_4OH or triethylamine salt seemed to antagonize possible interfering effects of sylanol residues.

Phosphoric acid was used instead of AcCOOH [127] to improve LTC recovery. Inadequate recovery of sulfidopeptide LTs was ascribed to interactions between the peptide sequence of LTs (this problem is less severe for LTB_4) and the column, leading to excessive retention. This problem was obviated by washing the chromatographic column with EDTA [31, 130]. The separation of sulfur-containing LTs is adversely affected by traces of metal cations in the stationary phase which can be removed by the EDTA. Moreover EDTA affords some protection against oxidation of LTC_4 due to acid or traces of heavy metals [131].

Some researchers improved the resolution by injecting the peaks isolated by RP-HPLC on the same column equilibrated with a water-enriched mobile phase, e.g. MeOH-water (60:40 or 40:30, v/v) acidified to pH 5.4 with AcCOOH [27,28].

The procedures described above allow purification of the single class of LTs. Systems consisting of a series of isocratic elutions of MeOH in water (e.g. from 48% to 100%) buffered at pH 5.8 have been reported as suitable for separating all three classes of AA metabolites in one single HPLC run [121]. Similar results were obtained by the gradient elution mode.

Isocratic mixtures and very shallow gradients of AcCN and 0.1% aqueous AcCOOH pH 3.7 have been used [132] to separate major PGs, LTs and HETEs in a single chromatographic procedure. AcCN was used because of its relatively good UV transparency below 200 nm for the detection of PGs. With regard to the use of different mobile phases in either the isocratic or gradient elution mode, the reader should refer to Table 3. This table shows that the solvent systems used are, with few exceptions, based on mixtures of alcohols with acidified water. Propanol has been used instead of MeOH [112]. A solvent system with neutral pH has been described [29] but in this HPLC system LTs coelute.

A LiChrosorb C-8 column, initial elution with buffer (0.123 M pyridine-0.4 M formic acid-0.001% thiodiglycol, pH 3.25) followed by a linear gradient between this buffer and 40% of the same buffer and 60% propanol has been described [124].

A trifluoroacetic acid (TFA) gradient with a constant concentration of AcCN in water has been used [134] for the analysis of peptido LTs. This system results in the elution of these products between HETEs and AA. For analysis of more complex mixtures containing LTs and cyclooxygenase products, linear gradient systems employing mobile phases composed of water-AcCN—MeOH and either TFA or phosphoric acid are suggested. Mobile phases containing mixtures of the above-mentioned acids do not require washing of the column with EDTA, as in the case of mobile phases containing AcCOOH. For preparative applications however, phosphoric acid is not volatile and TFA is quite a strong acid. Thus when the solvent is evaporated, the concentration of TFA, low as it may be, initially increases and the pH of the solution just prior to complete evaporation can be quite low. This difficulty can be reduced by the addition of a small amount of NH_4OH or triethylamine before removal of the solvent.

RP-HPLC and SP-HPLC have been used for the purification and fractionation of LTB_4 from various sources including peritoneal PMN [9,11,12,107,135–137] and peripheral blood PMN [13,16,39,61,63,101,103,127,138–141].

The solvent systems used for elution RP and SP columns are indicated in Table 4. The isocratic mixtures used with RP-ODS columns are generally based on aqueous MeOH (60%–80% by volume) mixtures acidified with AcCOOH (0.01%–0.1%).

The isocratic systems used with SP silica gel columns are based upon hexane-isopropanol mixtures (ratios 90:9 to 95:5, v/v) with AcCOOH (0.01%–0.02%).

A better resolution of LTB_4 from other AA metabolites can be achieved by gradient elution RP- or SP-HPLC [39,103]. However, the different isomers of LTB_4 cannot be completely separated in these conditions. Often the fractions obtained by RP-HPLC are methylated with diazomethane and the methyl esters of LTB_4 are separated by

Table 3. Purification of sulfidopeptide leukotrienes using HPLC

Source	Species	Mobile phase	Column	Ref.
Mastocytoma cell line	Mouse	MeOH/water (70:30) plus 0.02% AcCOOH	Polygosil 60-D_5C_{18}	26)
Mastocytoma cell line	Mouse	MeOH/water (69:31) plus 0.02% AcCOOH pH 5.7	RSIL C_{18} (preparative)	7, 126, 128)
		MeOH/water (65:35) plus 0.02% AcCOOH pH 5.7	Nucleosil C_{18} (analytical)	
Mastocytoma cell line	Mouse	MeOH/water (67:33) plus 0.02% H_3PO_4 pH 5.7	RP—C_{18}	127)
Mastocytoma cell line	Mouse	MeOH/water (67:33) plus 0.08% AcCOOH and EDTA 0.03%	Bondapak C_{18}	22)
RBL-1 cell line	Rat	MeOH/water (70:30) plus 0.1% AcCOOH pH 5.4	Polygosil C_{18}	27, 100)
		MeOH/water (60:40) plus 0.1% AcCOOH pH 5.4	Nucleosil C_{18}	
RBL-1 cell line	Rat	Gradient *n*-propanol in 5% aqueous AcCOOH	Bondapak *cis*	112)
RBL-1 cell line	Rat	MeOH/water (65:35) with 0.05% AcCOOH pH 5.4	Radial Pak C_{18}	28)
		MeOH/water (70:30) or (55:45)		
RBL-1 cell line	Rat	Gradient from 6% MeOH/0.6% *t*-pentanol/93.4% water pH 7.4 to 99.4% MeOH/0.6% *t*-pentanol	Varian C_{18} MCM Micropak	29)
RBL-1 cell line	Rat	MeOH/water (67:33) plus 0.02% AcCOOH pH 4.7	Nucleosil C_{18}	113)
RBL-1 cell line	Rat	MeOH/water (65:35) plus 0.01% AcCOOH pH 5.6	RP—C_{18}	115)
		MeOH/water (60:40) with 0.01% AcCOOH pH 5.6		
RBL-1 cell line	Rat	MeOH/water (68:32) with 0.05% AcCOOH pH 5.4	Nucleosil C_{18}	33)
RBL-1 cell line	Rat	AcCN/MeOH/0.1% AcCOOH (50:50:100) pH 5.6	Ultrasphere ODS	116)
Peritoneal cells	Rat	AcCN/MeOH/water/AcCOOH (33.6:5.4:61:1) pH 5.6	Novopak C_{18}	129)
Peritoneal cells	Rat	MeOH/water (68:32) with 0.1% AcCOOH pH 5.8	Ultrasphere C_{18}	70)
Peritoneal macrophages	Mouse	MeOH/water/AcCOOH (65:35:0.06) pH 5.4	Techsphere 5 C_{18}	125)
Peritoneal macrophages	Mouse	0.123 M pyridine/0.4 M HCOOH/0.001% thiodiglycol (Buffer 1) gradient from Buffer 1 to 40% Buffer 1/60% propanol	LiChrosorb C 8	124)
Alveolar macrophages	Mouse	MeOH/water/AcCOOH from 65% to 100% MeOH	Ultrasphere ODS	120)
Alveolar macrophages	Rat	MeOH/water (65:35) plus 0.01% AcCOOH pH 5.4	Nucleosil C_{18}	117, 118)
Alveolar macrophages	Rat	MeOH/0.1% aqueous AcCOOH pH 5.8 in different ratios from 48% MeOH to 100% MeOH	C_{18} Radial Pak	121)
Alveolar macrophages	Pig	MeOH/water/AcCOOH (65:34.9:0.1) pH 5.5	Bondapak C_{18}	119)
Peripheral blood PMN	Human Pig	THF/MeOH/water/AcCOOH (25:30:45:0.1) pH 5.5 with EDTA 0.1%	Nucleosil 5 C_{18}	34)

Peripheral blood eosinophils	Horse	MeOH/water (65:35) with 0.1% AcCOOH pH 3.9	Bondapak C_{18}	[30]
		MeOH/water (72:28) with 0.1% AcCOOH pH 3.9	Ultrasphere ODS C_{18}	
		MeOH/water (65:35) with 0.01% AcCOOH pH 5.7	Nucleosil C_{18}	
Peripheral blood eosinophils	Human	MeOH/water (70:30) with 0.1% AcCOOH pH 5.4	Nucleosil C_{18}	[36]
Lung mast cells	Human	70%–72% MeOH in 0.1% aqueous AcCOOH pH 4.0	Ultrasphere C_{18}	[133]
Bone marrow/mast cells	Mouse	MeOH/water/AcCOOH (65:34.9:0.1) pH 5.6	RP—C_{18}	[122]
Bone marrow/mast cells	Mouse	MeOH/water/AcCOOH (65:34.9:0.1) pH 5.6	Ultrasil C_{18}	[123]
		AcCN/water (40:60) pH 5.0	RP—C_{18}	

Table 4. Purification of LTB_4 using HPLC

Source	Species	Mobile phase	Column	Ref.
Peritoneal PMN	Rabbit	MeOH/water (70:30) plus 0.01% AcCOOH	μ Bondapak C_{18}	107)
		MeOH/water (75:25) plus 0.01% AcCOOH	Nucleosil C_{18}	
		Isopropyl alcohol/hexane (6:100)	μ Porasil	
		Isopropyl alcohol/hexane (4:100)	Nucleosil	
Peritoneal PMN	Rabbit	Hexane/isopropyl alcohol/MeOH/AcCOOH from (975:22:26:1) to (891:99:9:1)	LiChrosorb Si-100	9)
Peritoneal PMN	Guinea pig	MeOH/water (75:25) plus 0.01% or 0.1% AcCOOH	μ Bondapak C_{18}	135)
Peritoneal PMN	Rat	MeOH/water (75:25) plus 0.01% AcCOOH	Spherisorb 5 C_{18}	11, 12)
Peritoneal PMN	Rat	MeOH/water (75:25) plus 0.01% AcCOOH	Ultrasphere C_{18}	136)
		AcCN/water (65:35) plus 0.1% AcCOOH		
Peritoneal PMN	Rat	MeOH/water (65:35) plus 0.07% AcCOOH pH 5.5	RP—C_{18}	137)
Peripheral blood PMN	Human	MeOH/water (75:25) plus 0.01% AcCOOH	Nucleosil C_{18}	101)
Peripheral blood PMN	Human	MeOH/water (75:25) plus 0.01% AcCOOH	Polygosil C_{18}	13)
		Isopropyl alcohol/hexane (5:95) plus 0.01% AcCOOH	Nucleosil	
Peripheral blood PMN	Human	MeOH/water (70:30) plus 0.01% AcCOOH	Polygosil C_{18}	16)
		Isopropyl alcohol/hexane (5:95) plus 0.01% AcCOOH	Nucleosil 50-5	
Peripheral blood PMN	Human	MeOH/water (69:31) plus 0.02% AcCOOH	Zorbak C-8	138)
Peripheral blood PMN	Human	Gradient of isopropyl alcohol in hexane	Nucleosil 50-5	103)
Peripheral blood PMN	Human	Gradients of MeOH/AcCN (75:25) in MeOH/AcCN/water/AcCOOH (35:25:45:0.02)	Techsphere Ultra C_{18}	39)
		Hexane/isopropyl alcohol (100:12)	Techsphere Ultrasilica	
Peripheral blood PMN	Human	MeOH/water (67:33) plus 0.1% AcCOOH	Nucleosil C_{18}	139)
Peripheral blood PMN	Human	MeOH/water (70:30) plus 0.01% AcCOOH	RP—C_{18}	140)
		Hexane/isopropyl alcohol (90:10) plus 0.01% AcCOOH	SP silica	
Peripheral blood PMN	Human	Hexane/1-propanol/AcCOOH (1000:65:1)	μ Porasil	63)
Peripheral blood PMN	Human	MeOH/water (71:29) plus 0.02% AcCOOH pH 5.7	RP—C_{18}	127)
Peripheral blood PMN	Human	MeOH/AcCN/water/AcCOOH (35:25:45:0.02) to MeOH/AcCN (75:25)	Techsphere Ultra C_{18}	141)
Peripheral blood PMN	Human	MeOH/water (65:35), (70:30), (57:43) plus 0.1% AcCOOH	Altex ODS C_{18}	61)

Peripheral blood eosinophils	Horse	MeOH/water (65:35) plus 0.1% AcCOOH pH 3.9	μ Bondapak C_{18}	30)
		MeOH/water (72:28) plus 0.1% AcCOOH pH 3.9	Ultrasphere C_{18}	
		MeOH/water (65:35) plus 0.01% AcCOOH pH 5.7	Nucleosil C_{18}	
Peripheral blood leukocytes	Human	MeOH/water (70:30) plus 0.01% AcCOOH	Nucleosil 50-5 C_{18}	102)
		Hexane/isopropyl alcohol (95:5) plus 0.01% AcCOOH	Nucleosil 50-5	
Peripheral blood leukocytes	Human	Hexane/isopropyl alcohol/MeOH (85:10:5) plus 0.1% AcCOOH	Nucleosil 50-5	24)
Peripheral blood leukocytes	Pig	Gradient of isopropyl alcohol in hexane plus 0.1% AcCOOH	SP silica	142)
Erythrocytes	Human	MeOH/water (70:30) plus 0.1% AcCOOH	RP—C_{18}	38, 143)
		Hexane/isopropyl alcohol (200:10) plus 0.1% AcCOOH	SP silica	
Granulocytes	Human	Gradient AcCN in 0.1% aqueous AcCOOH	μ Bondapak C_{18}	144)
Peritoneal macrophages	Mouse	Hexane/EtOH (95:5) plus 0.1% AcCOOH	μ Porasil	17)
Alveolar macrophages	Rabbit	Gradient of MeOH in 0.05% aqueous AcCOOH	Ultrasphere ODS	21, 23)
Alveolar macrophages	Human	Gradient of AcCN in 0.058% aqueous H_3PO_4	Ultrasphere ODS	132)
Lung mast cells	Human	MeOH/water (70:30) plus 0.1% AcCOOH pH 4.0	Ultrasphere ODS	133)
Bone marrow-mast cells	Mouse	MeOH/water/AcCOOH (65:34.9:0.1) pH 5.6	C_{18} Ultrasil ODS	122)
Mastocytoma cell line	Mouse	MeOH/water (65:35) plus 0.02% AcCOOH	Nucleosil C_{18}	7, 126, 128)
		MeOH/water (69:31) plus 0.02% AcCOOH pH 5.7 and 4		
Mastocytoma cell line	Mouse	MeOH/water (67:33) plus 0.02% H_3PO_4 pH 5.7	RP—C_{18}	127)
PT-18 cell line	Mouse	Gradient of isopropyl alcohol in hexane	LiChrosorb Si-100	86)
RBL-1 cells	Rat	MeOH/water (75:25) plus 0.01% AcCOOH	RP—C_{18}	37)
RBL-1 cells	Rat	MeOH/water (68:32) plus 0.05% AcCOOH pH 4.7	Nucleosil C_{18}	114)
		Hexane/isopropyl alcohol (95:5) plus 0.02% AcCOOH	Nucleosil-50	

SP-HPLC purification using the above-mentioned mixtures hexane-isopropanol with or without AcCOOH [13, 16, 38, 102, 107, 114, 140]. Sometimes an additional final RP step is introduced. Actually the compounds converted to methyl esters are regenerated to the free acid form and again purified by RP-HPLC [39]. Mobile phases different from those mentioned above include AcCN instead of MeOH in the RP-HPLC [136] and the addition of MeOH to the SP solvent system, e.g. hexane-isopropanol-MeOH—AcCOOH [24], or EtOH instead of isopropanol, e.g. hexane-EtOH—AcCOOH [17]. Gradient programs from MeOH—AcCN-water-AcCOOH (35:25:45:0.02, v/v) to MeOH—AcCN (75:25, v/v) have also been used [141]. The addition of tetrahydrofuran (THF) has been suggested [34] to improve the resolution of LTB_4.

ω-oxidation products of LTB_4, 20-COOH—LTB_4 and 20-OH—LTB_4, have been purified from human leukocytes by following the same procedures as described for LTB_4 [102, 141]. Interaction of these products with the hydrophobic resin Nucleosil C18 is weaker than for LTB_4 because the oxidation of carbon 20 alters the physical properties of the molecule.

The epoxide intermediate LTA_4 is known to be an exceedingly labile substance, e.g. with mild aqueous acid, oxygen or free radicals. LTA_4 is first converted into a stable derivative, generally its methyl ester, and then purified by RP-HPLC [145] or SP-HPLC [146].

Alkaline mobile phase AcCN-0.01 M, pH 10 borate buffer (40:60, v/v) is suggested [145] to preserve LTA as an intact allylic epoxide, but this solvent system can erode the silica column; to minimize this it is advisable to use a precolumn or a saturation column.

3.3.2 Purification of Hydroxy and Hydroperoxy Arachidonic Acid Metabolites Using HPLC

Hydroxy- and hydroperoxy-derivatives of AA are separated from other AA metabolites by either RP-HPLC or SP-HPLC. This depends upon the fact that the class of hydroxy-derivatives spans a wide range of polarities and thus a clear distinction between HETE or HPETE and other AA metabolites cannot be obtained by simple RP-HPLC with isocratic elution. The increasing polarity order, as determined by RP-HPLC with acidified solvent systems, is: AA > HETEs—HPETEs > LT > PGs—TX and the elution order is consequently the reverse. Some cases of overlap among the classes exist however, due to the differences in the number of polar functional groups (hydroxyl, carboxyl) or in chromatographic conditions (pH and polarity of mobile phase, type of stationary phase). Thus, to obtain highly purified fractions it may sometimes be necessary to submit the fractions obtained by RP-HPLC to further SP-HPLC, or vice versa [147].

Fractions containing hydroxy- and hydroperoxy-derivatives can be purified from peptido LTs or by modifying the pH conditions in RP-HPLC or by using SP-HPLC that strongly absorbs the latter. Moreover, SP is generally better than RP at separating closely related positional and stereo isomers, but in some cases recoveries can be rather low.

The detection of HETEs or HPETEs can be performed by using the spectrophotometric UV detector set at around 235 nm for compounds having a conjugated double bond system as chromophore or set at a wavelength of 270 nm for a higher number of conjugated double bonds. The choice of the lower wavelength ensures the detection of all interesting compounds but has the disadvantage of a lower selectivity and a high level of interference. HPLC has been used for the purification and fractionation of hydroxy and hydroperoxy AA metabolites from various sources including peripheral blood leukocytes [146, 148–152], eosinophils [20], monocytes [67], platelets [147, 153, 154], keratinocytes [155], endothelial cells [156], smooth muscle cells [78], RBL-1 cells [10, 157] and peritoneal PMN [14].

The solvent systems for elution with RP and SP column are indicated in Table 5. They are like those used to fractionate LTB_4. Actually, as well as for LTB_4, the isocratic systems used with RP-ODS columns are generally based upon aqueous MeOH mixtures acidified with AcCOOH 0.01 %: for DHETEs MeOH 65 %, for their methyl esters 70 %, for HETEs and their methyl esters 75 %. Again the isocratic systems more frequently used with SP silica gel columns consist of mixtures of hexane-isopropanol (95:5 for DHETEs and 90:10 for THETEs) with or without AcCOOH (0.01 %–0.02 %). Gradient elution RP-HPLC [10, 20, 62, 78] or SP-HPLC [62, 148, 157] has been used with the above-mentioned solvents.

A better resolution of hydroxy and hydroperoxy derivatives of arachidonic acid has been achieved by using subsequentially RP and SP [10, 78, 147, 149–151, 157].

Lipoxin A and lipoxin B (5,6,15- and 5,14,15-trihydroxy-eicosatetraenoic acid) appear to be the major products of new series of compounds that contain four conjugated double bonds as a distinguishing feature and show a strong absorption at 301 nm. These compounds were isolated from human leukocytes exposed to 15 HPETE [69]. Purification was achieved by silicic acid column chromatography of the free acids, TLC of the methyl esters, and RP of the methyl esters on a Polygosil C18 column eluted with MeOH-water (70:30, v/v) and UV detector setting at 301 nm.

3.3.3 Purification of PGs and TXB_2 Using HPLC

The PGs and TXs, owing to the presence in their molecules of polar groups (carboxyl, carbonyl, hydroxyl in various combinations and number), are soluble in pure organic solvents as well as in aqueous organic mixtures, thus allowing both SP- and RP-HPLC separations.

In consequence of the type and number of conjugated double bonds, PGs and TXs display different UV absorption characteristics that can be used for HPLC detection. All the PGs and TXs share a common absorption in the short UV region (190–200 nm, maximum wavelength 192.5 nm) that may be used for general purpose detection. Molecules containing keto or conjugated double bonds, e.g. 15-keto-PGE_2 and PGB_2, can be more selectively detected at higher wavelengths (maxima at 217.2 and 278 nm, respectively).

The chromophore responsible for light absorption at 192.5 nm is probably the alkene group contained within the prostaglandins, with an average extinction coefficient of 15.800 as shown for various PGs in Table 6. The use of the wavelength of 192.5 nm for monitoring PGs however greatly reduces the choice of solvents that can be included in the mobile phase systems. A mixture of 0.017 M H_3PO_4—AcCN (67.2:32.8, v/v) has been successfully suggested by Terragno et al. [158]. Column life and, therefore, reproducibility was maintained with this solvent mixture because the system was constantly kept at pH 3.5, well within the recommended pH range of 2–7.5 for RP columns.

Table 5. Purification of hydroxy derivatives of arachidonic acid using HPLC

Source	Species	Mobile phase	Column	Ref.
Peritoneal PMN	Rat	Hexane/isopropyl alcohol (91:9)	μ Porasil	14)
Peripheral blood PMN	Human	MeOH/water (78:22) plus 0.01% AcCOOH	C_{18} LiChrosorb	4)
		Hexane/isopropyl alcohol/AcCOOH (983:16:1)	Ultrasil-Si	
Peripheral blood PMN	Human	Gradient hexane/isopropyl alcohol/AcCOOH	Lichrosorb 100	62)
		Gradient AcCN/MeOH/water pH 3.7	Zorbak ODS	
Peripheral blood eosinophils	Human	MeOH/water (65:35), (60:40), (75:25), (70:30) plus 0.01% AcCOOH	μ Bondapak C_{18}	20)
		Hexane/isopropyl alcohol (100:3) and (100:0.45)	μ Porasil	
		Gradient from chloroform/AcCOOH (500:2) to MeOH/AcCOOH (50:2)		
Peripheral blood monocytes	Human	Hexane/isopropyl alcohol (100:12)	μ Porasil	67)
		Gradient AcCN in aqueous AcCOOH	μ Bondapak	
Peripheral blood leukocytes	Guinea pig	MeOH/water (75:25) plus 0.01% AcCOOH	μ Radial Pak C_{18}	152)
	Human	MeOH/water (60:40) plus 0.05% AcCOOH		
Peripheral blood leukocytes	Human	Gradient isopropyl alcohol in hexane 0.1% AcCOOH	Chroprep Si 60	148)
	Pig	Isopropyl alcohol/hexane (4:96)	Hibar silica gel	
		MeOH/water (80:20)	Radial Pak C_{18}	
Peripheral blood leukocytes	Human	Hexane/isopropyl alcohol (95:5) plus 0.01% AcCOOH	Polygosil 50-10	149)
		MeOH/water (75:25) plus 0.01% AcCOOH	Nucleosil C_{18}	
		Hexane/isopropyl alcohol (97:3) plus 0.01% AcCOOH	Nucleosil 50-5	
Peripheral blood leukocytes	Human	MeOH/water (70:30) plus 0.01% AcCOOH	Polygosil C_{18}	150, 151)
		Hexane/isopropyl alcohol (92:8) plus 0.01% AcCOOH	Polygosil C60-10 SiO_2	
Peripheral blood leukocytes	Human	MeOH/water (60:40) plus 0.05% AcCOOH and NH_4OH pH 5.4	Whatman M9 ODS	146)
		Hexane/isopropyl alcohol (94:6) plus 0.1% AcCOOH	Radial Pak silica gel	
Plateletes	Human	Gradient AcCN in aqueous H_3PO_4	Zorbak ODS	153)
		0.4% isopropyl alcohol in hexane	Zorbak silica	
Platelets	Human	Hexane/isopropyl alcohol (97.5:2.5) plus 0.02% AcCOOH	Nucleosil 50 SiO_2	147)
		MeOH/water (70:30) plus 0.1% AcCOOH	RP—C_{18}	
Platelets	Rat	Hexane/isopropyl alcohol (98:2) plus 0.1% AcCOOH	Nucleosil 50-5	154)
Keratinocytes	Human	MeOH/water (77:23) plus 0.1% AcCOOH	Altex ODS	155)

Smooth muscle cells	Rat	Hexane/isopropyl alcohol/AcCOOH (991:8:1)	Micro Pak Si	[78]
		Gradient MeOH/0.01 % aqueous AcCOOH	μ Bondapak	
Endothelial cells	Human	MeOH/water (80:20) plus 0.01 % AcCOOH	Ultrasphere C_{18}	[156]
		MeOH/water (73.75:26.25) plus 0.01 % AcCOOH	RP—C_{18}	
RBL-1 cell line	Rat	Gradient MeOH in 0.05 % aqueous AcCOOH pH 4.7	Ultrasphere C_{18}	[10]
		Hexane/isopropyl alcohol (95:5) plus 0.01 % AcCOOH	Nucleosil 50	
RBL-1 cell line	Rat	Gradient AcCN in aqueous H_3PO_4	Spherisorb C-8	[157]
		Gradient hexane/isopropyl alcohol/AcCOOH from (96:4:0.01) to (90:10:0.01)	SP silica	

Table 6. Ultraviolet absorption data. Molar extinction coefficients (ε) and maximum wavelength λ_{max} (nm) in AcCN-aqueous phosphoric acid [158]

Compounds	ε	λ_{max} (nm)	Compounds	ε	λ_{max} (nm)
6-keto-PGF_{1a}	14.300	192.5	15-keto-PGE_2	10.060	217
6-keto-PGE_1	14.200	192.5	PGA_2	18.500	192.5
TXB_2	15.400	192.5	PGB_2	18.980	278
PGE_2	16.500	192.5	PGF_{2a}	17.100	192.5
PGE_1	15.800	192.5	PGF_{1a}	16.400	192.5
PGD_2	17.100	192.5			

The detection in the short UV region and also the low levels of these compounds, their instability and their rapid metabolic turnover have been considered as sources of trouble for HPLC separations. In this view derivatization with a UV or fluorescent chromophore improves the sensitivity and the specificity of detection and, at same time, the stability of the products during purification. Methods of derivatization, however, assume that the reactions are carried to completion and that losses do not occur. Fluorescent esters of PGE_2, F_2, D_2 and 6-keto-$PGF_{1\alpha}$ and TXB_2 have been formed by the addition of fluorogenic reagents such as 4-bromo-methyl-7-methoxy-coumarin [159] and 7-chloro-carbonyl-methoxy-4-methylcoumarin [160].
p-Nitrobenzyloximes of PG esters have been formed and purified using a reverse phase μ Bondapak C18 column with AcCN-water as the mobile phase and detection at 254 nm [161]. Derivatization of PGs enable their detection by UV absorbance and it has been very useful for the analysis of mixtures of synthetic PGs but not very useful for the analysis of these compounds from biological sources, due to the presence of unrelated compounds which absorb in the UV region. Moreover, the low volatility of these derivatives makes subsequent analysis difficult.

Also the use of "on line" radioactivity detectors is of limited utility for complex mixtures of PGs. This is because of the inherent peak broadening that occurs in these detectors to improve sensitivity, which would reduce resolution.

The conversion to methyl esters prior to purification has also been described, but this offers only the advantage of a better stability of these compounds. HPLC has been used for the purification and fractionation of cyclooxygenase products from various sources including endothelial cells [162–165], platelets [52, 159, 161, 166, 167] and macrophages [6, 120, 133]. Taking into account the possibility of using both SP- and RP-HPLC and of fractionating both underivatized and derivatized compounds it is not surprising to find many different separation procedures in the literature (see Table 7).

PGE_2, $PGF_{2\alpha}$ and their metabolites (13,14-dihydro-15-keto-PGE_2, 13,14-dihydro-15-keto-$PGF_{2\alpha}$) have been fractionated as methyl esters on a HC—ODS Silx column with a linear gradient of AcCN in water. The gradient started from 10% of AcCN in water, reaching 90% of AcCN in 15 min; a column rinse with pure AcCN for 15 min was necessary for the complete elution of the substances present in the sample [169]. The separation of PGI_2Na from 6-keto $PGF_{1\alpha}$ has been achieved on a Bondapak C18 column eluted with AcCN-water (20:80, v/v) buffered at pH = 9.3 with sodium borate. In fact, PGI_2 is an unstable compound, its vinyl ether functional group hydrolizes

Table 7. Purification of prostaglandins and thromboxane B_2 using HPLC

Source	Species	Mobile phase	Column	Ref.
Endothelial cells	Bovine	Gradient AcCN in aqueous H_3PO_4 pH 2.0	Altex ODS	162)
Endothelial cells	Bovine	Gradient AcCN in 0.06% aqueous H_3PO_4	Waters C_{18}	163)
Endothelial cells	Pig	Gradient from chloroform to MeOH/AcCOOH/chloroform (6:0.6:95.4)	SP-silica	164)
Endothelial cells		AcCN/water/toluene/HCOOH (23:76.7:0.2:0.1)	RP—C_{18}	
Smooth muscle sells fibroblasts	Pig	Water/AcCN/AcCOOH/benzene (385:115:0.5:1)	μ Bondapak C_{18}	165)
Lung mast cells	Human	Gradient AcCN in 0.1% aqueous AcCOOH pH 3.7	Ultrasphere C_{18}	133)
Alveolar macrophages	Mouse	MeOH/water/AcCOOH (65:34.9:0.1)	Ultrasphere ODS	6, 120)
Peritoneal macrophages		MeOH/AcCOOH (100:0.01)		
		Water/AcCN/benzene/AcCOOH (76.7:23:0.2:0.1)	Waters fatty acid	
Platelets	Human	AcCN/water (85:15) for PGs (55:45) for TXB_2	μ Bondapak C_{18}	161)
Platelets	Human	Gradient from chloroform/isooctane/MeOH (35:65:1.0) to chloroform/MeOH (100:1.35)	Varian CN-10 Micropak	159)
Platelets	Human	Gradient AcCN in aqueous H_3PO_4 pH 2	Spherisorb ODS	166)
Platelets	Human	Gradient AcCN in aqueous H_3PO_4	Zorbak ODS	52)
Platelets	Cat	Gradient from AcCN/water/benzene/AcCOOH (230:767:2:1) to (500:498:2:1)	Waters fatty acid ODS	167)
Kidney cell line	Dog	Gradient chloroform to 6% MeOH and 0.6% AcCOOH in chloroform	SP-silica	168)
		AcCN/water/benzene/AcCOOH (23:76.7:0.2:0.1)	Waters FA ODS	
Mastocytoma cell line	Mouse	Gradient AcCN in aqueous H_3PO_4	Ultrasphere ODS C_{18}	88)

rapidly in aqueous solutions by an acid-catalyzed process, but the increase of alkalinity retards this hydrolysis [170].

A procedure with a mixture of AcCN and aqueous 0.1% AcCOOH separates the major PGs, LTs, HETEs and AA [133]. Gradients of AcCN in aqueous phosphoric acid were used [52, 88, 162, 163, 166].

Mutual separation of cyclooxygenase and lipoxygenase products was carried out [6] on an Ultrasphere ODS column eluted with MeOH-water-AcCOOH (65:34.9:0.1, v/v). Separation of the individual cyclooxygenase products was on a Waters fatty acid analysis column with water-AcCN-benzene-AcCOOH (76.7:23:0.2:0.1, v/v) [6, 120].

Both cyclooxygenase and lipoxygenase products from platelets can be separated in one HPLC run with an initial mobile phase of water-AcCN-benzene-AcCOOH (767:230:2:1, v/v) then a linear gradient to water-AcCN-benzene-AcCOOH (498:500:2:1, v/v) [167].

PGE_2, $F_{2\alpha}$, 6-keto $PGF_{1\alpha}$ and TXB_2 from endothelial cells, smooth muscle cells and fibroblasts were purified on a μ C18 Bondapak column eluted with water-AcCN—AcCOOH-benzene (385:115:0.5:1, v/v) [165].

PGE_2, 6-keto-$PGF_{1\alpha}$ and $PGF_{2\alpha}$ were purified on a silicic acid column eluted with a linear gradient from chloroform to MeOH—AcCOOH-chloroform (6:0.6:95.4, v/v), and then, since 6-keto-$PGF_{1\alpha}$ and PGE_2 eluted together under these conditions, the peak containing these two substances was rechromatographed on a RP column eluted isocratically with AcCN-water-benzene-AcCOOH (23:76.7:0.2:0.1, v/v) [168, 171].

A second-step purification by means of a silver ion–loaded cation exchange column was suggested to separate underivatized PGs and HETEs. Retention times of the standards on this column are influenced by the polar group of the stationary phase reacting with the polar groups of the standards, and the interactions of the silver ions of the stationary phase with double bonds of the standards. The relative contribution of these two types of interactions could be changed by varying the composition of the mobile phase. Polar solvents containing low concentrations of AcCN allowed the maximum separation of the standard on the basis of the number of double bonds. As the polarity of the mobile phase was reduced or the concentration of AcCN increased; the selectivity of the stationary phase resembled that of SP on silica gel. This method will separate mixtures of closely related products and it is useful as second-step purification. The results obtained with a complex mixture can be difficult to interpret because structurally unrelated products of greatly different polarities (e.g. $PGF_{2\alpha}$ and 15 HETE) may have identical retention times [172].

4 Scale-up Considerations

AA metabolites are prepared in research laboratories in amounts of up to micrograms of highly purified compounds. The above-mentioned methods are in fact used for laboratory-scale preparations when structure and composition of metabolites present in cell culture fluids are to be investigated. In these methods the scaling-up, if required for clinical or pharmacological studies, may be facilitated thanks to the commercial availability of the equipment, which renders possible large-scale production of secreted

Table 8. Maximal yields of some arachidonic acid matabolites

Cellular source	Metabolite	Amount (μg per 10^7 cells)	Ref.
Murine mastocytoma cells	LTC_4	0.1–0.5	7, 121, 126, 127)
	LTB_4	0.02–0.1	
	PGD_2	0.3	
Tumor-derived murine macrophage cell lines (J774A.1 and WR19M.1)	PGD_2	1.0–2.0	88)
	PGE_2	0.2	
Mouse bone marrow–derived mast cells	LTC_4	0.9	122, 123)
	LTB_4	0.08	
	PGD_2	0.06	
Rat PMN	LTB_4	0.06–0.12	12, 137)
Porcine aortic endothelial cells	6-keto-$PGF_{1\alpha}$	3.3	164)
	PGE_2	2.8	
Human leukocytes	LTB_4	0.15–0.43	103, 148)
	5-HETE	0.4–3.5	
Human endothelial cells	PGI_2	1.4	75)
Horse eosinophils	LTB_4	0.09	30)

substances from mammalian cells as well as large preparative-scale chromatographic purification of cell extracts.

As far as the production of AA metabolites is concerned, when they cannot be obtained by chemical synthesis the conventional methods for growth of mammalian cells in vitro, e.g. roller bottles and tissue flask cultures, offer possibilities of production on the order of micrograms/day, as shown in Table 8.

These systems however do not provide surface area/volume ratios large enough for practical scale-up. Recently, microcarrier suspension cultures [173] and hollow fiber systems [174] have offered the possibility of building bioreactors where high-density cultures are mantained for a long time with continuous production of secreted biomolecules. In these cases the cell density may be increased by a factor of ten and the volume by a factor of one hundred. From a theoretical point of view this ought to allow the scaling up of the yield of the secreted products into the range of milligrams per day and for considerably longer periods (3–6 months).

A practical approach to the biosynthesis of large quantities of AA metabolites has been offered by the introduction of enzyme-immobilized techniques. Actually HPETEs have been produced by soybean lipoxygenase covalently coupled to agarose in amounts of 0.6 mg h^{-1} [175]. This method allows the continuous production of metabolites over many months and can be scaled-up without difficulty.

With regard to the purification of products obtained either by chemical or biological synthesis all the chromatographic techniques offer possibilities of scaling-up.

The TLC may be scaled-up from analytical to preparative dimensions by applying to the plates hundreds of milligrams of raw materials in the form of long bands. The maximum load of the plate can be obtained by increasing the thickness of the absorbent layer from 0.25 mm for analytical plates to 2 mm for preparative plates. This method is convenient for laboratory-scale preparations.

Open-column chromatography, which is essentially a laboratory-scale preparative technique, can be scaled up to purify a sample of a few grams without difficulty. Silica gel columns (100–120 μm, particles) of 50–60 cm length and 4–5 cm internal diameter allow the application of samples up to 10 g. The chromatographic process is not very efficient and the fractions obtained have to be further purified.

The introduction of specially designed columns allow so-called "flash chromatography", which is open-column chromatography carried out under pressure of an inert gas at high flow rates. Either SP or RP materials can be used in a process that is suitable for quickly separating various classes of AA metabolites.

The HPLC offers today the major possibilities in terms of costs, purity and rate of production [176]. Large preparative columns and opportune chromatographic apparatus are commercially available. By means of computer-assisted recycling of the peaks of different columns these specialized chromatographs allow the treatment of hundreds of grams per day of raw material with an efficiency comparable to analytical HPLC [177]. The columns used in this equipment have a size of 10 × 50 cm and a particle dimension of 10–30 μm.

Both SP and RP columns are available and virtually all the analytical procedures can be translated into pilot or industrial dimensions. However this possibility is not so effective since SP materials cost 20 times less then the corresponding bonded phases.

5 Acknowledgement

The authors are very grateful to Mrs. Liliana Francardi for her careful and patient typing of the manuscript.

6 List of Abbreviations

AcCOOH	acetic acid
AcCN	acetonitrile
AA	arachidonic acid
DHETE	dihydroxy-eicosatetraenoic acid
EtOH	ethanol
EtOAc	ethylacetate
GC/MS	gas chromatography/mass spectrometry
HPLC	high-performance liquid chromatography
HPETE	hydroperoxy-eicosatetra-enoic acid
HETE	hydroxy-eicosatetraenoic acid
LT	leukotriene
MeOH	methanol
ODS	octadecyl silyl
PMN	polymorphonuclear leuko-cytes
PG	prostaglandin
RBL	rat basophilic leukemia
RP	reverse-phase
RIA	radioimmunoassay
SRS	slow reacting substance
SP	straight-phase
TFA	trifluoroacetic acid
THETE	trihydroxy-eicosatetraenoic acid
THF	tetrahydrofuran
TLC	thin layer chromatography
TX	thromboxane

7 References

1. Folch, J., Lees, M., Sloane-Stanley, G. H.: J. Biol. Chem. *226*, 497 (1957)
2. Bligh, E. G., Dyer, W. J.: Can. J. Biochem. Physiol. *37*, 911 (1959)
3. Saunders, R. D., Horrocks, L. A.: Anal. Biochem. *143*, 71 (1984)
4. Soberman, R. J., Harper, T. W., Betteridge, D., Lewis, R. A., Austen, K. F.: J. Biol. Chem. *260*, 4508 (1985)
5. Scott, W. A., Pawlowski, N. A., Andreach, M., Cohn, Z. A.: J. Exp. Med. *155*, 535 (1982)
6. Rouzer, C. A., Scott, W. A., Griffith, O. W., Hamill, A. L., Cohn, Z. A.: Proc. Natl. Acad. Sci. USA *79*, 1621 (1982)
7. Westcott, J. Y., Murphy, R. C.: Prostaglandins *26*, 221 (1983)
8. Powell, W. S.: Methods Enzymol. *86*, 467 (1982)
9. Vanderhock, J. Y., Bryant, R. W., Bailey, J. M.: J. Biol. Chem. *255*, 10064 (1980)
10. Wei, Y., Evans, R. W., Morrison, A. R., Sprecher, H., Jakschik, B. A.: Prostaglandins *29*, 537 (1985)
11. Ford-Hutchinson, A. W., Bray, M. A., Doig, M. V., Shipley, M. E., Smith, M. J. H.: Nature (London) *286*, 264 (1980)
12. Ford-Hutchinson, A. W., Bray, M. A., Cunningham, F. M., Davidson, E. M., Smith, M. J. H.: Prostaglandins *21*, 143 (1981)
13. Claesson, H. E., Lundberg, U., Malmsten, C.: Biochem. Biophys. Res. Commun. *99*, 1230 (1981)
14. Ham, E. A., Soderman, D. D., Zanetti, M. E., Dougherty, H. W., McCauley, E., Kuehl, F. A., Jr.: Proc. Natl. Acad. Sci. USA *80*, 4349 (1983)
15. Hansson, G., Lindgren, J. A., Dahlén, S. E., Hedqvist, P., Samuelsson, B.: FEBS Lett. *130*, 107 (1981)
16. Lindgren, J. A., Hansson, G., Samuelsson, B.: FEBS Lett. *128*, 329 (1981)
17. Humes, J. L., Sadowski, S., Galavage, M., Goldenberg, M., Subers, E., Bonney, R. J., Kuehl, F. A., Jr.: J. Biol. Chem. *257*, 1591 (1982)
18. Maas, R. L., Turk, J., Oates, J. A., Brash, A. R.: J. Biol. Chem. *257*, 7056 (1982)
19. Jakschik, B. A., Lee, L. H., Shuffer, G., Parker, C. W.: Prostaglandins *16*, 733 (1978)
20. Turk, J., Maas, R. L., Brash, A. R., Roberts II, L. J., Oates, J. A.: J. Biol. Chem. *257*, 7068 (1982)
21. Hsueh, W., Sun, F. F.: Biochem. Biophys. Res. Commun. *106*, 1085 (1982)
22. Musch, M. W., Bryant, R. W., Coscolluela, C., Myers, R. F., Siegel, M. J.: Prostaglandins *29*, 405 (1985)
23. Hsueh, W., Sun, F. F., Henderson, S.: Biochim. Biophys. Acta *835*, 92 (1985)
24. Camp, R. D. R., Fincham, N. J.: Br. J. Pharmacol. *85*, 837 (1985)
25. Gréen, K., Hamberg, M., Samuelsson, B., Frölich, J. C.: Adv. Prostaglandin and Thromboxane Res. *5*, 25 (1978)
26. Murphy, R. C., Hammarström, S., Samuelsson, B.: Proc. Natl. Acad. Sci. USA *76*, 4275 (1979)
27. Örning, L., Hammarström, S., Samuelsson, B.: ibid. *77*, 2014 (1980)
28. Sok, D. E., Pai, J. K., Atrache, V., Sih, C. J.: ibid. *77*, 6481 (1980)
29. Parker, C. W., Falkenhein, S. F., Huber, M. M.: Prostaglandins *20*, 863 (1980)
30. Jörg, A., Henderson, W. R., Murphy, R. C., Klebanoff, S. J.: J. Exp. Med. *155*, 390 (1982)
31. Metz, S. A., Hall, M. E., Harper, T. W., Murphy, R. C.: J. Chromatogr. *233*, 193 (1982)
32. Granström, E., Kindahl, H.: Adv. Prostaglandin and Thromboxane Res. *5*, 119 (1978)
33. Jakschik, B. A., Kuo, C. G.: Adv. Prostaglandin, Thromboxane and Leukotriene Res. *11*, 141 (1983)
34. Verhagen, J., Walstra, P., Veldink, G. A., Vliegenthart, J. F. G.: Prostaglandins, Leukotrienes and Medicine *13*, 15 (1984)
35. Jubiz, W., Nolan, G., Kaltenborn, K. C.: J. Liquid Chromatogr. *8*, 1519 (1985)
36. Shaw, R. J., Walsh, G. M., Cromwell, O., Moqbel, R., Spry, C. J. F., Kay, A. B.: Nature (London) *316*, 150 (1985)
37. Ford-Hutchinson, A. W., Piper, P. J., Samhoun, M. N.: Br. J. Pharmacol. *76*, 215 (1982)
38. Fitzpatrick, F., Liggett, W., McGee, J., Bunting, S., Morton, D., Samuelsson, B.: J. Biol. Chem. *259*, 11403 (1984)
39. Shak, S., Goldstein, I. M.: ibid. *259*, 10181 (1984)
40. Bedetti, C., Cantafora, A.: Purification of arachidonic acid metabolites from cell cultures by

octadecyl reversed phase cartridges, in: Production & Exploitation of Existing and New Animal Cell Substrates (eds. Spier, E. R., Griffiths, B., Hennessen, W.), p. 331, S. Karger, Basel 1985
41. Gréen, K., Samuelsson, B.: J. Lipid Res. *5*, 117 (1964)
42. Hamberg, M., Samuelsson, B.: J. Biol. Chem. *241*, 257 (1966)
43. Nijkamp, F. P., Moncada, S., White, H. L., Vane, J. R.: Eur. J. Pharmacol. *44*, 179 (1977)
44. Hamberg, M., Samuelsson, B.: Proc. Natl. Acad. Sci. USA *71*, 3400 (1974)
45. Shah, N. T., Karpen, C. W., Panganamala, R. V.: Thromb. Res. *23*, 225 (1981)
46. Harris, D. N., Phillips, M. B., Michel, I. M., Goldenberg, H. J., Heikes, J. E., Sprague, P. W., Antonaccio, M. J.: Prostaglandins *22*, 295 (1981)
47. Sun, F. F., McGuire, J. C., Morton, D. R., Pike, J. E., Sprecher, H., Kunau, W. H.: Prostaglandins *21*, 333 (1981)
48. Killackey, J. J., Killackey, B. A., Philp, R. B.: Prostaglandins, Leukotrienes and Medicine *9*, 9 (1982)
49. Boukhchache, D., Lagarde, M.: Biochim. Biophys. Acta *713*, 386 (1982)
50. Fisher, S., Weber, P. C.: Biochem. Biophys. Res. Commun. *116*, 1091 (1983)
51. Aharony, D., Smith, J. B., Silver, M. J.: Biochim. Biophys. Acta *718*, 193 (1982)
52. Evans, R. W., Sprecher, H.: Prostaglandins *29*, 431 (1985)
53. Hashimoto, Y., Naito, C., Teramoto, T., Kato, H., Kinoshita, M., Kawamura, M., Hayashi, H., Oka, H.: Biochem. Biophys. Res. Commun. *130*, 781 (1985)
54. Haurand, M., Ullrich, V.: J. Biol. Chem. *260*, 15059 (1985)
55. Bailey, J. M., Bryant, R. W., Feinmark, S. J., Makheja, A. N.: Prostaglandins *13*, 479 (1977)
56. Sun, F. F., Chapman, J. P., McGuire, J. C.: ibid. *14*, 1055 (1977)
57. Bryant, R. W., Bailey, J. M.: Biochem. Biophys. Res. Commun. *92*, 268 (1980)
58. Mencia-Huerta, J. M., Hadji, L., Benveniste, J.: J. Clin. Invest. *68*, 1586 (1981)
59. Marcus, A. J., Broekman, M. J., Safier, L. B., Ullman, H. L., Islam, N.: Biochem. Biophys. Res. Commun. *109*, 130 (1982)
60. Punnonen, K., Uotila, P.: Prostaglandins, Leukotrienes and Medicine *15*, 177 (1984)
61. Soberman, R. J., Harper, T. W., Murphy, R. C., Austen, K. F.: Proc. Natl. Acad. Sci. USA *82*, 2292 (1985)
62. Vanderhoek, J. K., Karmin, M. T., Ekborg, S. L.: J. Biol. Chem. *260*, 15482 (1985)
63. Billah, M. M., Bryant, R. W., Siegel, M. I.: ibid. *260*, 6899 (1985)
64. Borgeat, P., Samuelsson, B.: J. Biol. Chem. *254*, 2643 (1979)
65. Siegel, M. I., McConnell, R. T., Bonser, R. W., Cuatrecasas, P.: Prostaglandins *21*, 123 (1981)
66. Kimura, Y., Okuda, H., Arichi, S.: Biochim. Biophys. Acta *834*, 275 (1985)
67. Goldyne, M. E., Burrish, G. F., Oliver, C.: Prostaglandins *30*, 77 (1985)
68. Yoshimoto, T., Miyamoto, Y., Ochi, K., Yamamoto, S.: Biochim. Biophys. Acta *713*, 638 (1982)
69. Serhan, C. N., Hamberg, M., Samuelsson, B.: Proc. Natl. Acad. Sci. USA *81*, 5335 (1984)
70. Dobson, P., Aharony, D., Krell, R. D.: Res. Commun. Chem. Pathol. Pharmacol. *42*, 3 (1983)
71. Wood, J. N., Coote, P. R., Rhodes, J.: FEBS Lett. *174*, 143 (1984)
72. de Maroussem, D., Pipy, B., Beraud, M., Derache, P., Mathieu, J. R.: Biochim. Biophys. Acta *834*, 8 (1985)
73. Hsueh, W., Desai, U., Gonzalez Crussi, F., Lamb, R., Chu, A.: Nature (London) *290*, 710 (1981)
74. Punnonen, K., Uotila, P., Mäntylä, E.: Res. Commun. Chem. Pathol. Pharmacol. *44*, 367 (1984)
75. Baenziger, N. L., Becherer, P. R., Majerus, P. W.: Cell *16*, 967 (1979)
76. Larrue, J., Rigaud, M., Daret, D., Demond, J., Durand, J., Bricaud, H.: Nature (London) *285*, 480 (1980)
77. Larrue, J., Leroux, C., Daret, D., Bricand, H.: Biochim. Biophys. Acta *710*, 257 (1982)
78. Nakao, J., Koshihara, Y., Ito, H., Murota, S. I., Chang, W. C.: Life Sci. *37*, 1435 (1985)
79. Willems, C., De Groot, P. G., Pool, G. A., Gonsalvez, M. S., Van Aken, W. G., Van Mourik, J. A.: Biochim. Biophys. Acta *713*, 581 (1982)
80. Thomas, J. M. F., Hullin, F., Chap, H., Douste-Blazy, L.: FEBS Lett. *176*, 202 (1984)
81. Gorodetsky, J. S., Hla, T. T., Bailey, J. M.: Pharmacol. Res. Commun. *17*, 447 (1985)
82. Goldsmith, J. C., Needleman, S. W.: Prostaglandins *24*, 173 (1982)
83. Coene, M. C., Van Hove, C., Claeys, M., Herman, A. G.: Biochim. Biophys. Acta *710*, 437 (1982)
84. Harvey, J., Holgate, S. T., Peters, B. J., Robinson, C., Walker, J. R.: Br. J. Pharmacol. *86*, 417 (1985)

85. Shoam, H., Razin, E.: Biochim. Biophys. Acta *837*, 1 (1985)
86. Vanderhoek, J. Y., Tare, N. S., Bailey, J. M., Goldstein, A. L., Pluznik, D. H.: J. Biol. Chem. *257*, 12191 (1982)
87. Vanderhoek, J. Y., Pluznik, D. H.: Biochim. Biophys. Acta *837*, 119 (1985)
88. McGuire, J. C., Richard, K. A., Sun, F. F., Tracey, D. E.: Prostaglandins *30*, 949 (1985)
89. Jakschik, B. A., Lee, L. H.: Nature (London) *287*, 51 (1980)
90. Steinhoff, M. M., Lee, L. H., Jakschik, B. A.: Biochim. Biophys. Acta *618*, 28 (1980)
91. Casey, F. B., Appleby, B. J., Buck, D. C.: Prostaglandins *25*, 1 (1983)
92. Falkenhein, S. F., MacDonald, H., Huber, M. M., Del Kock, Parker, C. W.: J. Immunol. *125*, 163 (1980)
93. Furukawa, M., Yoshimoto, T., Ochi, K., Yamamoto, S.: Biochim. Biophys. Acta *795*, 458 (1984)
94. Hamasaki, Y., Tai, H. H.: ibid. *834*, 37 (1985)
95. Ubatuba, F. B.: J. Chromatogr. *161*, 165 (1978)
96. Goswami, S. K., Kinsella, J. E.: J. Chromatogr. *209*, 334 (1981)
97. Goswami, S. K., Kinsella, J. E.: Lipids *16*, 759 (1981)
98. Samuelsson, B.: J. Biol. Chem. *238*, 3229 (1963)
99. Almè, B., Hansson, G.: Prostaglandins *15*, 199 (1978)
100. Örning, L., Bernström, K., Hammarström, S.: Eur. J. Biochem. *120*, 41 (1981)
101. Borgeat, P., Samuelsson, B.: Proc. Natl. Acad. Sci. USA *76*, 2148 (1979)
102. Claesson, H. E., Lindgren, J. A., Gustafsson, B.: Biochim. Biophys. Acta *836*, 361 (1985)
103. Maclouf, J., Fruteau de Laclos, B., Borgeat, P.: Proc. Natl. Acad. Sci. USA *79*, 6042 (1982)
104. Parker, C. W., Jakschik, B. A., Huber, M. A., Falkenhein, S. F.: Biochem. Biophys. Res. Commun. *89*, 1186 (1979)
105. Maas, R. L., Brash, A. R., Oates, J. A.: Novel leukotrienes and lipoxygenase product from rat mononuclear cells, in: SRS-A and Leukotrienes (ed. Piper, P. J.), p. 151, New York, Wiley 1981
106. Yecies, L. D., Parker, C. W., Watanabe, S.: Life Sci. *25*, 1909 (1979)
107. Borgeat, P., Samuelsson, B.: J. Biol. Chem. *254*, 7865 (1979)
108. Lewis, R. A., Austen, K. F., Drazen, J. M., Clark, D. A., Marfat, A., Corey, E. J.: Proc. Natl. Acad. Sci. USA *77*, 3710 (1980)
109. Corey, E. J., Barton, A. E., Clark, D. A.: J. Am. Chem. Soc. *102*, 4278 (1980)
110. Corey, E. J., Marfat, A., Goto, G., Brion, F.: J. Am. Chem. Soc. *102*, 7984 (1980)
111. Corey, E. J., Clark, D. A., Goto, G., Marfat, A., Mioskowski, C., Samuelsson, B., Hammarström, S.: J. Am. Chem. Soc. *102*, 1436 and 3663 (1980)
112. Morris, H. R., Taylor, G. W., Piper, P. J., Samhoun, M. N., Tippins, J. R.: Prostaglandins *19*, 185 (1980)
113. Jakschik, B. A., Harper, T., Murphy, R. C.: J. Biol. Chem. *257*, 5346 (1982)
114. Jakschik, B. A., Kuo, C. G.: Prostaglandins *25*, 767 (1983)
115. Bach, M. K., Brashler, J. R., Morton, D. R.: Arch. Biochem. Biophys. *230*, 455 (1984)
116. Yoshimoto, T., Soberman, R. J., Lewis, R. A., Austen, K. F.: Proc. Natl. Acad. Sci. USA *82*, 8399 (1985)
117. Rankin, J. A., Hitchock, M., Merrill, W., Back, M. K., Brashler, J. R., Askenase, P. W.: Nature (London) *297*, 329 (1982)
118. Rankin, J. A., Hitchock, M., Merrill, W., Huang, S. S., Brashler, J. R., Bach, M. K., Askenase, P. W.: J. Immunol. *132*, 1993 (1984)
119. Paterson, N. A. M.: Am. Rev. Respir. Dis. *129*, 274 (1984)
120. Rouzer, C. A., Scott, W. A., Hamill, A. L., Cohn, Z. A.: J. Exp. Med. *155*, 720 (1982)
121. Henke, D. C., Kouzan, S., Eling, T. E.: Anal. Biochem. *140*, 87 (1984)
122. Razin, E., Mencia-Huerta, J. M., Lewis, R. A., Corey, E. J., Austen, K. F.: Proc. Natl. Acad. Sci. USA *79*, 4665 (1982)
123. Mencia-Huerta, J. M., Razin, E., Ringel, E. W., Corey, E. J., Hoover, D. M., Austen, K. F., Lewis, R. A.: J. Immunol. *130*, 1885 (1983)
124. Rouzer, C. A., Scott, W. A., Cohn, Z. A., Blackburn, P., Manning, J. M.: Proc. Natl. Acad. Sci. USA *77*, 4928 (1980)
125. Osborne, D. J., Peters, B. J., Meade, C. J.: Prostaglandins *26*, 817 (1983)
126. Murphy, R. C., Mathews, W. R.: Methods Enzymol. *86*, 409 (1982)
127. Westcott, J. Y., Murphy, R. C.: Biochim. Biophys. Acta *833*, 267 (1985)
128. Mathews, W. R., Rokach, J., Murphy, R. C.: Anal. Biochem. *118*, 96 (1981)

129. Abe, M., Kawazoe, Y., Shigematsu, N.: Anal. Biochem. *144*, 417 (1985)
130. Anderson, F. S., Westcott, J. Y., Zirrolli, J. A., Murphy, R. C.: Anal. Chem. *55*, 1837 (1983)
131. Westcott, J. Y., Clay, K. L., Murphy, R. C.: J. Allergy Clin. Immunol. *74*, 363 (1984)
132. Peters, S. P., Schulman, E. S., Lin, M. C., Hayes, E. C., Lichtenstein, L. M.: J. Immunol. Methods *64*, 335 (1983)
133. Peters, S. P., MacGlashan, D. W., Jr., Schulman, E. S., Schleimer, R. P., Hayes, E. C., Rokach, J., Adkinson, N. F., Jr., Lichtenstein, L. M.: J. Immunol. *132*, 1972 (1984)
134. Powell, W. S.: Anal. Biochem. *148*, 59 (1985)
135. Bokoch, G. M., Reed, P. W.: J. Biol. Chem. *256*, 4156 (1981)
136. Stenson, W. F., Prescott, S. M., Sprecher, H.: J. Biol. Chem. *259*, 11784 (1984)
137. Evans, J. F., Nathaniel, D. J., Zamboni, R. J., Ford-Hutchinson, A. W.: J. Biol. Chem. *260*, 10966 (1985)
138. Salmon, J. A., Simmons, P. M., Palmer, R. M. J.: FEBS Lett. *146*, 18 (1982)
139. Clancy, R. M., Dahinden, C. A., Hugli, T. E.: Proc. Natl. Acad. Sci. USA *81*, 5729 (1984)
140. Serhan, C. N., Lundberg, U., Weissmann, G., Samuelsson, B.: Prostaglandins *27*, 563 (1984)
141. Shak, S., Reich, N. O., Goldstein, I. M., Ortiz de Montellano, P. R.: J. Biol. Chem. *260*, 13023 (1985)
142. Naccache, P. H., Molski, T. F. P., Becker, E. L., Borgeat, P., Picard, S., Vallerand, P., Sha'afi, R.: J. Biol. Chem. *257*, 8608 (1982)
143. McGee, J., Fitzpatrick, F.: ibid. *260*, 12832 (1985)
144. Luderer, J. R., Riley, D. L., Demers, L. M.: J. Chromatogr. *273*, 402 (1983)
145. Fitzpatrick, F. A., Morton, D. R., Wynalda, M. A.: J. Biol. Chem. *257*, 4680 (1982)
146. Sok, D. E., Han, C. O., Shieh, W. R., Zhou, B. N., Sih, C. J.: Biochem. Biophys. Res. Commun. *104*, 1363 (1982)
147. Wong, P. Y. K., Westlund, P., Hamberg, M., Granström, E., Chao, P. H. W., Samuelsson, B.: J. Biol. Chem. *260*, 9162 (1985)
148. Borgeat, P., Picard, S., Drapeau, J., Vallerand, P.: Lipids *17*, 676 (1982)
149. Radmark, O., Serhan, C., Hamberg, M., Lundberg, U., Ennis, M. D., Bundy, G. L., Oglesby, T. D., Aristoff, P. A., Harrison, A. W., Slomp, G., Scahill, T. A., Weissmann, G., Samuelsson, B.: J. Biol. Chem. *259*, 13011 (1984)
150. Lundberg, U., Radmark, O., Malmsten, C., Samuelsson, B.: FEBS Lett. *126*, 127 (1981)
151. Jubiz, W., Radmark, O., Lindgren, J. A., Malmsten, C., Samuelsson, B.: Biochem. Biophys. Res. Commun. *99*, 976 (1981)
152. Sok, D. E., Han, C. Q., Pai, J. K., Sih, C. J.: Biochem. Biophys. Res. Commun. *107*, 101 (1982)
153. Van Rollins, M., Horrocks, L., Sprecher, H.: Biochem. Biophys. Acta *833*, 272 (1985)
154. Pace-Asciak, C. R., Granström, E., Samuelsson, B.: J. Biol. Chem. *258*, 6835 (1983)
155. Burrall, B. A., Wintroub, B. U., Goetzl, E. J.: Biochem. Biophys. Res. Commun. *133*, 208 (1985)
156. Gorman, R. R., Oglesby, T. D., Bundy, G. L., Hopkins, N. K.: Circulation *72*, 708 (1985)
157. Goetze, A. M., Fayer, L., Bouska, J., Bornemeier, D., Carter, G. W.: Prostaglandins *29*, 689 (1985)
158. Terragno, A., Rydzik, R., Terragno, N. A.: Prostaglandins *21*, 101 (1981)
159. Turk, J., Weiss, S. J., Davis, J. E., Needleman, P.: Prostaglandins *16*, 291 (1978)
160. Karlsson, K. E., Wiesler, D., Alasandro, M., Novotny, M.: Anal. Chem. *57*, 229 (1985)
161. Fitzpatrick, F. A., Wynalda, M. A., Kaiser, D. G.: Anal. Chem. *49*, 1032 (1977)
162. Friedman, M., Madden, M. C., Saunders, D. S., Gammon, K., White, G. C., Kwoch, L.: Prostaglandins *30*, 1069 (1985)
163. Miller, D. K., Sadowski, S., Soderman, D. D., Kuehl, F. A., Jr.: J. Biol. Chem. *260*, 1006 (1985)
164. Whorton, A. R., Young, S. L., Data, J. L., Barchowsky, A., Kent, R. S.: Biochim. Biophys. Acta *712*, 79 (1982)
165. Ody, C., Seillan, C., Russo-Marie, F.: Biochim. Biophys. Acta *712*, 103 (1982)
166. Van Rollins, M., Aveldano, M. I., Sprecher, H. W., Horrocks, L. A.: Methods Enzymol. *86*, 518 (1982)
167. Ellis, E. F., Cockrell, C. S.: Prostaglandins, Leukotrienes and Medicine *13*, 153 (1984)
168. Alam, I., Ohuchi, K., Levine, L.: Anal. Biochem. *93*, 339 (1979)
169. Conte, D., Laguzzi, G., Boniforti, L., Cantafora, A., Di Silverio, F., Latino, C., Lalloni, G., Mesolella, V., Isidori, A.: INSERM *91*, 89 (1979)
170. Fitzpatrick, F. A.: The stability of eicosanoids: analytical consequences, in: Prostaglandins,

Prostacyclin, and Thromboxanes measurement (eds. Boeynaems, J. M., Herman, A. G.), p. 189, M. Nijhoff, The Hague 1980
171. Whorton, A. R., Smigel, M., Oates, J. A., Frölich, J. C.: Biochim. Biophys. Acta *529*, 176 (1978)
172. Powell, W. S.: Anal. Biochem. *115*, 267 (1981)
173. Microcarrier cell culture, principles & methods, Pharmacia Fine Chemicals, p. 34, Uppsala 1981
174. Ku, K., Kuo, M. J., Delente, J. Wildi, B. S., Feder, J.: Biotechnol. Bioeng. *23*, 79 (1981)
175. Laakso, S.: Lipids, *17*, 667 (1982)
176. Mann, A. F.: Int. Biotech. Lab. *4*, 28 (1986)
177. Colin, H., Lowy, G., Cazes, J.: Int. Biotech. Lab. *4*, 30 (1986)

Carbon Dioxide Transfer in Biochemical Reactors

Chester S. Ho*, Mark D. Smith and John F. Shanahan
Biochemical Engineering Laboratory, Department of Chemical Engineering, State University of New York at Buffalo, Buffalo, New York 14260/U.S.A.

1 Solubility of Carbon Dioxide in Aqueous Media 84
2 Diffusion of Carbon Dioxide in Aqueous Media 88
2.1 Diffusion and Diffusivities 88
2.2 Diffusion in the Presence of Ions 91
2.3 The Relation Between Mass Transfer and Diffusivity 92
2.4 Controlling Resistance and the Mass Transfer Coefficient 94
2.5 Carbon Dioxide Transfer in Microbial Systems 96
2.6 Conclusions 97
3 Reactions Involving CO_2 97
4 Desorption of Carbon Dioxide in Agitated Liquids 101
5 The Effect of Carbon Dioxide on Various Microbial Processes 107
6 The Effect of CO_2 on Mycelial Morphology 114
6.1 Conclusions for Applications 122
7 Conclusions 122
8 Acknowledgement 122
9 Nomenclature 122
10 References 124

Carbon dioxide is a major product of the oxidation of carbohydrates during cell metabolism. In essence, the oxidation of substrates, such as carbohydrates to CO_2 and H_2O, is the basis for aerobic forms of life. As such, carbon dioxide is a key component in life processes.

Carbon dioxide is not only a product, but also necessary for the growth of many microorganisms. The depletion of the TCA cycle pool constituents under certain conditions of growth necessitates carbon dioxide fixation to make up the deficiency. Carbon dioxide has also been proven to have a variety of effects on microbial morphology, spore germination, and cell growth and product formation.

A state-of-the-art review of the literature relating to carbon dioxide transfer in submerged biochemical reactors is presented. Emphasis is laid upon absorption and desorption of carbon dioxide in aqueous media. Considerable effort is devoted to discussions of physiological effects of carbon dioxide on microbial activities and carbon dioxide reactions in aqueous media.

* Author to whom correspondence should be addressed.

Advances in Biochemical Engineering/
Biotechnology, Vol. 35
Managing Editor: A. Fiechter

1 Solubility of Carbon Dioxide in Aqueous Media

Determination of the solubility of gaseous components is necessary to biological processes in establishing mass balances, calculating microbial yield coefficients, and in the determination of volumetric mass transfer coefficients. Hence, solubility information is essential for successful design and scale-up of bioprocesses [1]. For many years, the importance of the solubility of oxygen under various fermentation conditions has been recognized. Carbon dioxide, on the other hand, has largely been ignored and few sources of information are available in the literature. A brief review on the solubility of carbon dioxide in aqueous media will be presented.

There are many different ways of expressing gas solubility. A popular form is that of Henry's law, i.e.,

$$C_g = H \times C^* \tag{1}$$

Henry's law applies as long as the concentration of dissolved gas is small and the temperature and pressure are far removed from the critical temperature and pressure. Other forms which are often used to express gas solubilities include the Bunsen coefficient, the absorption coefficient, the Koenen coefficient and weight solubility. A full description of each coefficient can be found in Ref. [2].

The Bunsen coefficient, α, may be related to the Henry's law constant by Eq. (2) and is frequently used in the literature.

$$\alpha = (273.15/T) \times H \tag{2}$$

In general, there are two different approaches for the determination of gas solubilities in aqueous media. One method is founded on a theoretical basis and relies on mathematical formulations in order to predict the solubilities under a variety of circumstances. The other method is experimental in nature and results in empirical formulations in order to predict the solubilities of gases in solutions.

Gas solubility in a solution depends on many factors, of which temperature and pressure are two important ones for typical submerged bioprocesses. In most biological solutions there are far more complexities involved in affecting the solubilities of gases that may be present in the system. For example, the concentrations of a variety of dissolved inorganic and organic substances that are always present in the media play a crucial role in either decreasing or increasing the solubility of a gas.

Electrolytes have been known to reduce the solubility of carbon dioxide as well as other gases in what is referred to as a "salting-out" effect. The salting-out effect has been treated by Long and McDevit and Konnik and is discussed in Ref. [2]. On the other hand, a majority of nonelectrolytes have similar effects on gas solubilities. Efforts have been made by many researchers in order to develop an accurate and physically consistent model that will predict gas solubilities in a solution composed of just one electrolyte or mixed electrolytes [2,8,70].

Sechenov [4] described this decreased solubility effect for most electrolytes by the empirical equation:

$$\log(\alpha_0/\alpha_{el}) = K \times C_{el} \tag{3}$$

where α_0 is the Bunsen coefficient of water, α_{el} is the coefficient of a salt solution having a concentration of C_{el}, and K is the Sechenov constant which value depends on the kind of gas, the temperature, and the respective salts. At electrolyte concentrations beyond a critical value the predicted solubilities from Eq. (3) become too low. This critical concentration may be higher than 7 mol L^{-1} (as is with $NaNO_3$) or less than 1 mol L^{-1} (as is the case with acids showing a degree of dissociation deviating from Eq. (3). Sechenov constants calculated from the solubility measurement at high electrolyte concentrations may be too small [2,4]. This should be a consideration when using any of the semi-empirical methods presented in this review.

Schumpe et al. [2,3] presented a model that not only applies to single electrolyte solutions, but also to mixed electrolyte solutions. The following equation is applicable:

$$\log(\alpha_0/\alpha_{el}) = \sum_i H_i \times I_i \tag{4}$$

where

$$I_i = 0.5 \sum_i c_i \times Z_i^2 \tag{5}$$

I_i is the ionic strength attributable to each ion, i, and H_i is a parameter specific to the present, the ions, and the temperature. Values of H_i are given for CO_2 at 25 °C in Table 1 [2,8]. Fig. 1 gives a graphical comparison of the methods of Onda et al. [7] and Schumpe et al. [2,3] for CO_2 at 25 °C in the solutions of H_2SO_4 + NaCl and HCl + Na_2SO_4. The modified model given by Schumpe Eq. (4) is in good agreement with the experimental data given by Quicker et al. [1].

In addition to ions or salts, many organic solutes can be found in culture media. For example, one of the culture media for *Penicillium chrysogenum* is given in Table 2.

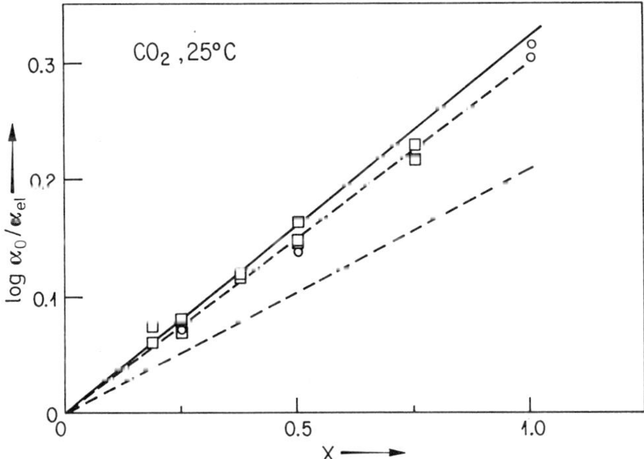

Fig. 1. CO_2 solubilities at 25 °C in solutions of (H_2SO_4 + NaCl) and (HCl + Na_2SO_4). Predictions of models and experimental values. (———) indicates data from Schumpe et al. [2,3] (-----------) indicates the prediction range of Onda et al. [7]

Table 1. Parameters H_i (for anions and cations in l mole^{-1}) for the model suggested by Schumpe et al. [2,8]

Cation	$H_i(CO_2)$ 25 °C	Anion	$H_i(CO_2)$ 25 °C
H^+	−0.319	Cl^-	0.339
Li^+	−0.178	Br^-	0.324
Na^+	−0.130	I^-	0.309
K^+	−0.196	NO_3^-	0.293
Rb^+	−0.217	HSO_4^-	0.436
Cs^+	−0.243	HSO_3^-	0.400
NH_4^+	−0.252	SO_4^{2-}	0.213
Mg^{2+}	−0.078	$S_2O_3^{2-}$	0.211
Ca^{2+}	−0.073		
Ba^{2+}	−0.064		
Mn^{2-}	−0.084		
Fe^{2+}	−0.078		
Cu^{2+}	−0.090		
Al^{3+}	−0.059		

The dependence of gas solubility on the concentrations of such chemical species has been described by the simple equation for the salting-out effect.

$$\log(\alpha_0/\alpha) = K \times C_n \tag{6}$$

Here C_n reflects the concentration of the organic solute and K is an empirical constant. This dependency may also be described by a linear relationship, namely:

$$\alpha = \alpha_0 * (1 - m * C_n) \tag{7}$$

The variable m in this case is a parameter determined by experiments. There are systems, however, that are simply not governed by such relationships. As a result, these expressions must be utilized with caution.

Some organic compounds can enhance the solubility of certain gases such as carbon dioxide and oxygen in solutions. Low weight chain alcohols have been known

Table 2. Culture media used for *Penicillium chrysogenum*

Corn steep liquor	75.00 g L^{-1}
Glucose	40.00 g L^{-1}
Yeast extract	16.00 g L^{-1}
Soybean oil	4.00 g L^{-1}
$(NH_4)_2SO_4$	3.60 g L^{-1}
KH_2PO_4	2.40 g L^{-1}
$MgSO_4$	2.30 g L^{-1}
$ZnSO_4$	0.01 g L^{-1}

to increase the solubility of some gases in water. Unfortunately, there is little of this type of data available for carbon dioxide in the literature.

Sugars such as glucose, lactose, and sucrose are all used as carbon sources for various microorganisms. These sugars have an effect on the solubility of the gases present. It has been shown that the concentration dependence of sugar on gas solubility is specific to the type of sugar involved. This dependence may not be represented by a single correlation. Prediction of the solubility of CO_2 can be made for lactose and glucose over a wide range of concentrations and with sucrose up to a concentration of 200.0 g L^{-1} by utilizing Eq. (6) [1]. To be able to predict the solubility over a wide range of sucrose concentrations, Eq. (7) must be used. Graphical representations are given in Fig. 2.

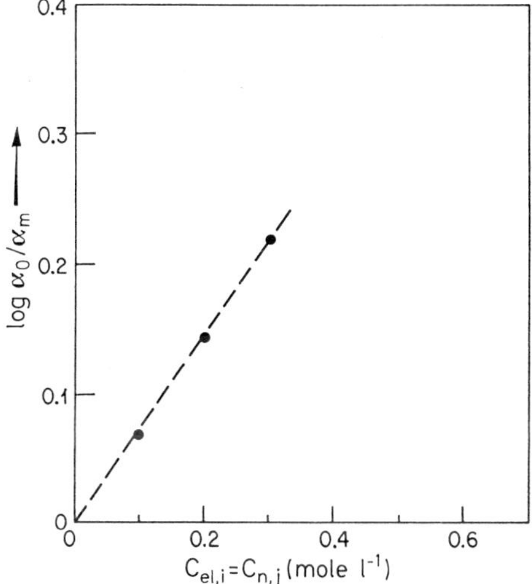

Fig. 2. Solubility reduction of CO_2: (---------) Prediction; (●) sucrose + glucose + $CuSO_4$ + $MgSO_4$

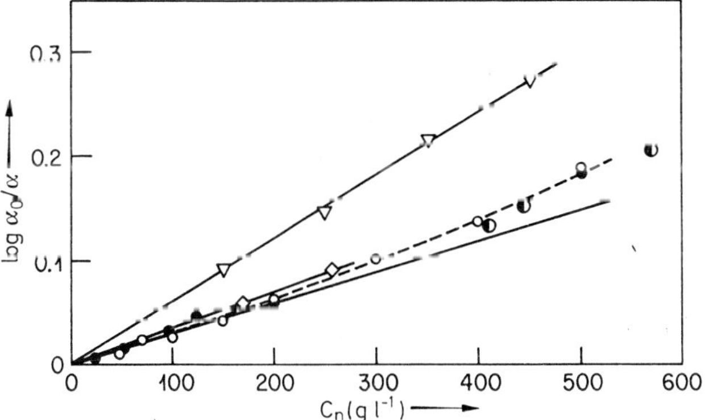

Fig. 3. CO_2 solubilities in solutions of glucose (▽), Koch; lactose (◇), Quicker et al. [1]; and sucrose (●), Findlay and Shen [6]; (◐), Koch, (○), Quicker et al. [1]

The question remains as to how to estimate gas solubilities in complex media composed mostly of sugars and salts. It was reasoned by Schumpe et al. that a logical extension of the previous work would lead to a log additive approach. In other words, the individual contributions of each of the species were found to be log additive with respect to the overall solubility reduction of the entire system [1,2]. Fig. 3 shows a system prediction of sucrose, glucose, $CuSO_4$, and $MgSO_4$ as compared to experimental data [1,70].

To further complicate the prediction of gas solubility, many living microorganisms are introduced into the already complex medium. The growth of these microorganisms leads to changes in medium composition due to conversion of the carbon sources and depletion of nutrients such as nitrogen sources or salts to biosynthetic products. Most cultivations involve the addition of acids or bases for pH control and antifoam agents to control foaming. In addition, the microorganisms are respiring and producing products. All of these conditions contribute to a changing and complex gas solubility.

2 Diffusion of Carbon Dioxide in Aqueous Media

The primary goal of biochemical engineering is to provide cells with an optimal environment for growth and product formation. In striving to achieve this goal, one important consideration is the mass transfer of essential elements to the living microorganisms. Information on mass transfer is used to optimize the cell environment to increase product yield. In studying mass transfer in biotechnological processes, several factors must be carefully considered. Clearly, if a vessel is mechanically agitated there are convective mechanisms involved. It is obvious that the hydrodynamics of solutions will affect the mass transfer.

The effect of agitation and aeration will be described in a stepwise approach. First the simple concepts such as the basic mass transfer descriptions involving the diffusion of solute A in solvent B will be treated, followed by the diffusion of electrolytes in aqueous solutions and a brief review of various models describing the agitated systems. This part of the chapter will summarize the development of these mass transfer models. Next, the mass transfer of component A in the presence of some general bulk liquid phase containing microorganisms will be considered. The objective is to show that the hydrodynamics of an agitated system does have a marked influence on the mass transfer of essential materials to living cells.

2.1 Diffusion and Diffusivities

In discussions of diffusion, the implication that there is some kind of force pushing molecules in the direction of the concentration gradient tends to be misleading. In actuality, molecular transport of a solute in a liquid is a consequence of random thermal motions. The overall result is a redistribution of the solute in a direction from regions of higher concentrations to regions of lower concentrations.

In the case of a stagnant liquid or film, there are no velocity components associated with the fluid. As a result, only the molecular diffusion becomes important in the mass transport of a solute. This situation may be expressed mathematically by Eq. (8):

$$\frac{\partial c}{\partial t} = D_{AB} \frac{\partial^2 c}{\partial x^2} \tag{8}$$

This represents an unsteady state problem meaning the transport via molecular diffusion varies with time. In general, the concentration of a solute varies both with time

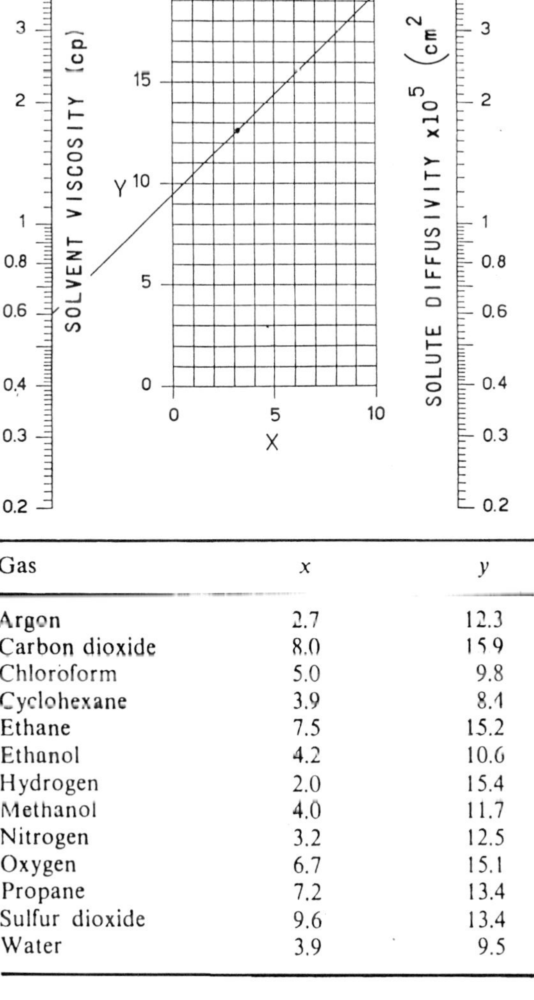

Gas	x	y
Argon	2.7	12.3
Carbon dioxide	8.0	15.9
Chloroform	5.0	9.8
Cyclohexane	3.9	8.1
Ethane	7.5	15.2
Ethanol	4.2	10.6
Hydrogen	2.0	15.4
Methanol	4.0	11.7
Nitrogen	3.2	12.5
Oxygen	6.7	15.1
Propane	7.2	13.4
Sulfur dioxide	9.6	13.4
Water	3.9	9.5

Fig. 4. A nomograph

and position in the liquid phase. It is important to note that the molecular diffusivity of the solute in the solvent, D_{AB}, may depend on position, concentration, and time. In the development of Eq. (8), it has been assumed that the diffusivity is independent of these variables. Details of the derivation of Eq. (8) may be found in the literature [5, 9, 10].

Estimating the solute diffusivity in liquid systems is a hardship often encountered in academic and industrial research efforts. Wilke and Chang have presented a semi-empirical method for diffusivity estimation.

$$D_{AB} = 7.4 \times 10^{-8} \frac{T(\chi M_B)^{1/2}}{\mu V_A^{0.6}} \tag{9}$$

where χ is an association factor, M_b is molecular weight of the solvent, T is temperature in K, μ is viscosity of the solvent in cp, and V_A is molecular volume of the solute at the normal boiling point in $cm^3\ g^{-1}\ mol^{-1}$. A disadvantage of this correlation is the need for experimental determination of χ for associated systems.

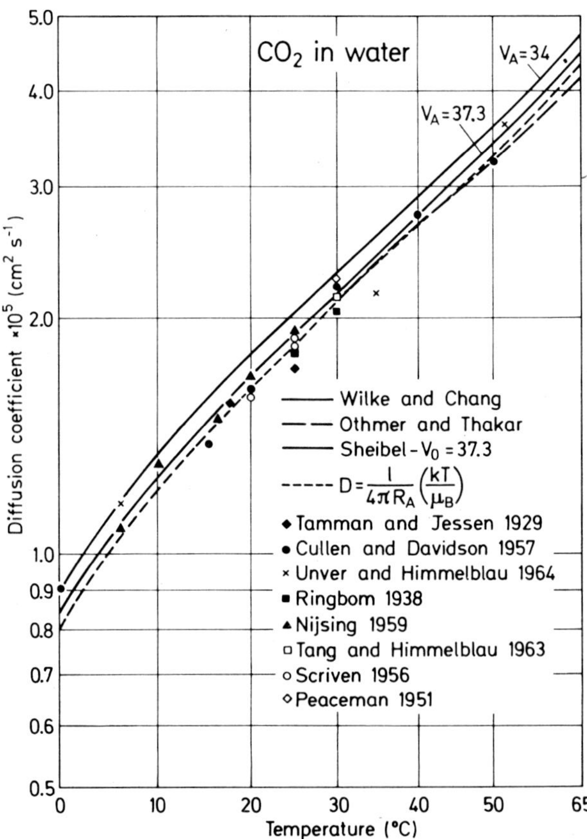

Fig. 5. Comparison of predicted and experimental diffusivities: carbon dioxide in water

Table 3. Selected values of the diffusivity of carbon dioxides in water, based upon experimental data at 1 bar

Gas	Liquid	Temp. (°C)	Diffusivity $\times 10^5$ ($cm^2\ s^{-1}$)
CO_2	H_2O	6.5	1.15
		10.0	1.30
		15.0	1.45
		20.0	1.75
		25.0	1.95
		30.0	2.25
		40.0	2.30
		50.0	3.23
		65.0	4.30

Other correlations exist but these also have pitfalls which tend to limit their use. Another method of obtaining diffusivity data is from a nomograph (see Fig. 4). This method is strictly based on an empirical relationship that states that D_{AB} is proportional to the viscosity of component B, usually the solvent, to some power. Generally, nomographs are less and cover a more limited range of variables than do correlations.

Sometimes diffusivity data are directly available in the literature. In Table 3 and Fig. 5 the diffusivity of carbon dioxide in water are presented at different temperatures. Interpolating to conditions that are within the range presented can be accomplished by plotting the data. To extrapolate to conditions that are outside those presented, it is best to plot the data and to fit them with a correlation such as the Wilke-Chang equation. This provides a qualitative guide for the extrapolation [10].

2.2 Diffusion in the Presence of Ions

When carbon dioxide is in aqueous solution, it undergoes a second order chemical reaction. This results in the production of carbonic acid and carbonate ions. The diffusion of the dissolved CO_2 is now in an electrolyte solution which may change its diffusivity. In real production processes there are many salts present in the solution and these contribute to the diverse electrolyte concentration in the medium. Danckwerts gives a brief discussion on estimating diffusivities of solutes in the presence of one electrolyte or many electrolytes [5].

Various researchers have studied the diffusivities of CO_2 in different solutions as well as in the presence of electrolytes. Davies et al. [12] have measured the diffusivities of carbon dioxide in a number of organic liquids. They have found that the diffusivity varies as $\mu^{-0.5}$ rather than μ^{-1} as in the Wilke-Chang equation. These researchers covered a 50-fold range of viscosities as compared to the 5-fold range covered by Wilke and Chang. Ratcliff and Holdcroft [13] have compared the diffusivities in a number of aqueous solutions and conclude that the diffusivity varies roughly as $\mu^{-0.637}$; while Nijsing et al. [14] have found the diffusivity of carbon dioxide in various solutions varying as $\mu^{-0.85}$.

2.3 The Relation Between Mass Transfer and Diffusivity

Several models have been developed by various researchers in order to describe the physical behavior of gas absorption in agitated liquids. These models lead to quantitative descriptions for the physical mass transfer and provide a physical insight to the factors affecting mass transfer. A brief presentation of three models that deal with mass transfer through a gas-liquid interface will be given [10].

The film model proposed by Whitman pictures a stagnant film of thickness δ at the surface of the liquid adjacent to the gas. Imagine that the gas concentration decreases from a value C* at the film surface to C at the bulk liquid-liquid film interface. If it is assumed that the bulk liquid is agitated and no convection exists within the film itself, then the only mode of transport will be by molecular diffusion.

$$\bar{R} = \frac{D_A(C^* - C)}{\delta} \tag{10}$$

The mass transfer coefficient may be defined as:

$$k_1 = \frac{D_A}{\delta} \tag{11}$$

where δ depends on many factors including liquid agitation, and the physical properties of the fluid. The model is rather simplistic in that it has such a drastic change or discontinuity so close to the gas-liquid interface. In addition, the assumption that this film zone has a uniform thickness, δ, seems equally idealized. Still, the model illustrates that a gas must enter the liquid by molecular diffusion before it can be augmented by convective mechanisms [5].

A second group of models involve the still surface approach. Here, as a distance is traversed from the gas-liquid interface, transport by turbulent mechanisms become more important until it becomes the dominating mechanism. Although this model avoids the sudden discontinuity present in the film model in that there is a progressive transition, there is a complication. The transfer coefficient that can be derived from the model contains two parameters relating to the hydrodynamic conditions instead of the single parameter, δ. A thorough description of this model can be found in Ref. [5].

A third category of model involves the surface-renewal approach. The foundation of this type of model relies on the replacement of discrete liquid elements at the gas-liquid interface from the interior of the liquid. This recycling of fluid elements may be induced by agitation creating a bulk motion of the fluid.

In the case of absorption of a gas, the fluid elements act as if they were infinitely deep and quiescent. Controlling the absorption rate of the gas is the exposure time of each element. Higbie's model makes a simplifying assumption that every liquid element has the same exposure time to the gas. After the passage of this time period, the liquid elements will be replaced by other elements each having the same exposure time as their predecessors. This exposure time period is very dependent on the hydrodynamic properties of the liquid and it is also the only parameter required to account for the hydrodynamic effect on the transfer coefficient, k_1.

Danckwerts modified the penetration theory proposed by Higbie by adding a complication. Instead of having equal exposure times, it was suggested that the chance of replacement of fluid elements at the surface is independent of the time the surface element was exposed to the gas. The result is an age distribution of fluid elements and is represented by Eq. (12) [5].

$$\varphi = \int_0^\infty s e^{-s\theta} \, d\theta \tag{12}$$

The rate of absorption per unit area of surface exposed for time is now an instantaneous quantity and not an average quantity as it was in the Higbie treatment. The quantity s represents the fraction of an area of surface which is replaced with fresh liquid in unit time. By summing up all of the instantaneous rates over the entire spectrum of the fluid elements present, the average rate of absorption may be obtained.

$$\bar{R} = s \int_0^\infty R \, e^{-s\theta} \, d\theta \tag{13}$$

Equation (13) may be solved if the instantaneous rate of absorbtion is known. Danckwerts gives this rate in Eq. (14) [5].

$$R = (C^* - \bar{C}) \sqrt{D_A/\pi\theta} \tag{14}$$

With this equation, the average rate can be solved as

$$\bar{R} = (C^* - \bar{C}) \sqrt{D_A s} \tag{15}$$

where

$$k_l = \sqrt{D_A s} \tag{16}$$

Here, the mass transfer coefficient is proportional to the square root of the diffusivity and the quantity s. The quantity s accounts for the hydrodynamic properties of the system.

Several other models have been proposed, each with their advantages and complications, and can be found listed in Refs. [5, 15] and [17]. One of the major difficulties with these models is the presence of quantities such as δ and s which can not be measured by conventional techniques. Still, they do succeed in providing a physical insight into the complex problem of describing an agitated system and the interfacial mass transfer involved.

There are empirical equations in the literature that relate the mass transfer coefficient, K_L, and the specific interfacial area, a, to various operating variables. Rushton's result for agitated tanks is given in Eq. (17) [19].

$$K_l = K \left(\frac{d^2 N \varrho}{\mu}\right)^b \left(\frac{\varrho D_A}{\mu}\right)^a \left(\frac{D_A}{d}\right) \tag{17}$$

Calderbank also gives a correlation for the interfacial area as [20]:

$$a = c\left[\frac{(P_g/V)^{0.4}\,\varrho^{0.2}}{\sigma^{0.6}}\right]\left(\frac{V_S}{V_T}\right)^{0.5} \tag{18}$$

Ionic species, as mentioned earlier, affect the molecular diffusivity. The mass transfer coefficient may also be expected to vary with different ion concentration graduents.

In Newtonian fluids, a physical absorption result for oxygen is presented in Ref. [18].

$$K_L a = \lambda\left(\frac{P_g}{V_L}\right)^n\left(\frac{F_g}{A}\right)^m \frac{\varrho_1^{.533} D_{02}^{2/3}}{\sigma^{0.6}\mu^{1/3}} \tag{19}$$

where V_L is the liquid volume, P_g is the power input during aeration, and A is the reactor cross sectional area perpendicular to the gas flow rate. The quantities λ, n, and m versus ionic strength are given by an empirical fit which can also be located in Ref. [18].

Many biochemical molecules present in the aqueous system possess hydrophobic and hydrophilic ends. These molecules tend to concentrate at gas-liquid or liquid-liquid interfaces and are referred to as surface active agents. This accumulation creates foaming in aerated vessels. One may initially guess that this increase of interfacial area would enhance the mass transfer as it does with the bubble aeration. There are counteracting mechanisms, however, that significantly decrease K_L. This adverse effect is thought to be the result of a combination of two mechanisms or just one of the mechanisms working alone [70].

The first mechanism to be considered proposes that the ease of liquid movement near the interface is decreased. Several mass transfer theories, such as the renewal theories which rely on small fluid elements being exchanged from the bulk liquid to the interface, predict that a loss of movement would result in reduced rates of exchange and a lower mass transfer coefficient. The second mechanism involves the formation of an actual film resistive to mass transfer. This newly formed barrier is enough to depart from the presumed equilibrium in the plane of the interface. In practice, various anti-foam agents have been employed to reduce the foam formation.

2.4 Controlling Resistance and the Mass Transfer Coefficient

Understanding gas transfer in biosystems is of considerable importance for several reasons. The product of an aerobic microbial population may be severely reduced if there is not enough dissolved oxygen present. This same result may occur if a substantial build-up of an inhibitory waste product such as carbon dioxide is allowed to accumulate beyond a certain maximum tolerable concentration. Oxygen transfer has been well categorized and explained in the literature. Its effect on growth and product yield has been quantitatively expressed in many biosystems. Carbon dioxide, on the other hand, has received little attention in the past. One of the reasons behind this was a lack of reliable sensors capable of readily assessing the in situ carbon dioxide partial pressure in the complex culture medium [16]. Before attention is directed at any one specific gas, the transfer coefficient must be discussed.

Experimentally, the gas-liquid mass transfer while considering only the physical absorption in a mixture is characterized by the following equations.

$$\text{flux of the gas} = k_g(C_g - C_g^*) \tag{20}$$

$$= k_l(C^* - C) \tag{21}$$

Equation (20) represents the gas side mass transfer and Eq. (21) represents the liquid-side mass transfer. Both the gas-side and liquid-side media offer resistances to the transfer of gas molecules, and one of the phases may control the rate of transfer. The concentrations at the gas-liquid interface are given by Eq. (22).

$$HC^* = C_g \tag{22}$$

Here, H is the Henry's law constant, C^* is the interfacial liquid composition in equilibrium with the concentration of the bulk gas concentration, and C_g is the bulk gas concentration. These concentrations are normally not accessible in mass transfer measurements. Defining the overall mass transfer coefficient as

$$\text{flux} = K_L(C^* - C) \tag{23}$$

allows the relationship between the overall mass transfer coefficient and the individual gas side and liquid side coefficients to be derived.

$$\frac{I}{K_L} = \frac{I}{k_l} + \frac{I}{Hk_g} \tag{24}$$

Clearly, if H is very large, as is the case for sparingly soluble species, the liquid-side mass transfer coefficient is the main contributor to the overall coefficient.

$$K_L \approx k_l \tag{25}$$

Essentially, all the resistance to transfer lies on the liquid side. This will be assumed in all future discussions involving the O_2 and CO_2 transfers.

The actual uptake rate by the entire growth process is given by the following equation,

$$Q_i = K_La(C^* - C) \tag{26}$$

The quantity a is the total interfacial area per unit volume of liquid in the bioreactor. The uptake rate is actually only a local value and a sum should be made over the entire volume of the bioreactor for an average uptake rate, $\bar{Q}$. Often in agitated vessels, the assumption is made that the local uptake rate is equivalent to the average uptake rate, thus,

$$Q_i = \bar{Q} \tag{27}$$

This assumption is only valid if the hydrodynamic conditions and the liquid concentrations, C* and C, are uniform throughout the vessel. Obviously, there are possibilities to be in a transition regime between mild agitation and strong agitation depending on the system involved. However one must always bear in mind that assuming an isotropic character in the fermentation vessels may not always be a sound approximation [70].

2.5 Carbon Dioxide Transfer in Microbial Systems

It is becoming very important to understand the dynamics of carbon dioxide accumulation in the culture medium during the process course. For example, the rate of CO_2 transfer will determine to a great extent the magnitude of the product inhibition in the submerged processes. Considering the fact that CO_2 is a respiratory waste product, low CO_2 transfer rates, especially in the desorption process, will result in a substantial build up of this inhibitory species in the medium when the cellular carbon dioxide production is of a higher magnitude. The sequence of transfer steps is depicted in Fig. 6.

Applying Eq. (26), the transfer rate of CO_2 across the gas liquid interface can be described by the following expression:

$$-r_{CO_2} = K_L a(P_{ai} - P_a) = K_L a(C_a^* - C_a) \tag{28}$$

Viable cells produce CO_2 at a specific rate generally related to the process conditions. It should be emphasized that carbon dioxide is a polar molecule and can be rapidly transported across cell membranes to the cell-liquid interface. Once in the aqueous phase, CO_2 is hydrolyzed to HCO_3^- and H^+. For the entire system, the rate of accumulation of carbon dioxide can be described as

$$\frac{d(C_a)}{dt} = K_1(C_a) - K_{-1}(HCO_3^-) - R'X' + K_L a(C_a^* - C_{ai}) \tag{29}$$

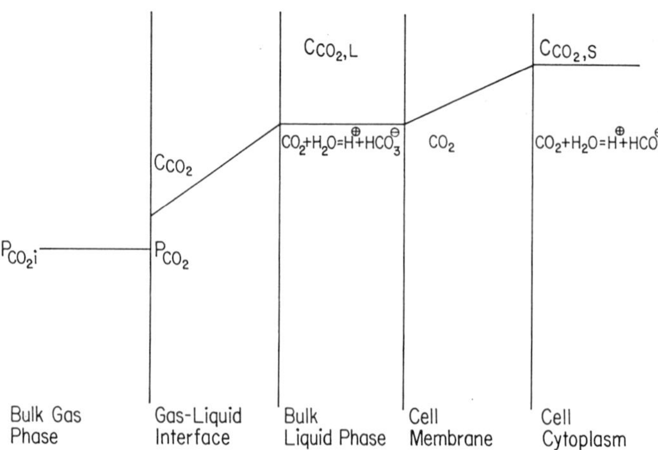

Fig. 6. Transfer steps of carbon dioxide from a cell to the gas bubble

where R′ is the specific carbon dioxide production rate and X′ is the viable cell mass per unit volume of medium. Therefore, the accumulation of CO_2 in the medium is a complex and dynamic process that is dependent on the viable cell concentration, pH, the mass transfer coefficient, and the specific CO_2 production rate [16, 70].

2.6 Conclusions

One of the major goals of the attempts to quantitatively describe the diffusivity of carbon dioxide and the relationship between diffusivity and the mass transfer coefficient (K_La) of CO_2 is to provide the basis for a sound scale-up and control for practical of commercially viable bioreactors. The foundation of such scale-up design must be firm and rely on reasonable and reproducible data and correlations. Unfortunately, diffusity data for CO_2 in salt solutions, glucose solutions and complex media is not available nor are relations between the diffusion coefficient and the mass transfer coefficient for CO_2 developed. Clearly more work is essential if CO_2 is to be used as a criterion for scale up or a control parameter.

In the future, the range of application of data for carbon dioxide must be carefully scrutinized. Predictions must be substantiated by experiment before "general" correlations can be developed. Due to the very nature of the bioprocesses and the associated complexities, especially with the physical and reactive properties of dissolved CO_2, correlations from basic data deserve marked caution and may be system dependent. Nevertheless, research on diffusivity and mass transfer coefficients of CO_2 should prove useful in determining the proper way to operate a bioprocess.

3 Reactions Involving CO_2

Invariably, various chemical reactions take place in complex media. Naturally, the respiration and metabolic processes of a cell involve an extremely complicated network of reactive pathways allowing its existence and growth. There are, however, reactions occurring in the external environment of the microorganism. Often, these reactions are complicated by virtue of the quantity of diverse chemical species present in most media. Such an abundance of different chemicals is vital for the normal functioning of a microorganism. It is of no surprise that the ability to study the reactions of a sole chemical compound is very difficult. If the product yield or the selectivity is to be altered by manipulation of the nutrients comprising the spent medium, then it becomes advantageous for the experimenter to have a knowledge of the possible reactions consuming various elements of interest.

The most thoroughly studied reaction is that of carbon dioxide and water. More than 99% of the dissolved CO_2 in water exists as a dispersed gas. Less than 1% exists as carbonic acid, H_2CO_3, which partially dissociates to yield hydrogen ions, H^+, bicarbonate ions, HCO_3^-, and a very small quantity of carbonate ions, $CO_3^=$ [21].

The first step in the gas-liquid system is the dissolution of carbon dioxide into water, i.e.,

$$CO_2(g) + H_2O \rightleftharpoons CO_2(d) + H_2O \quad (30)$$

Dissolved carbon dioxide then hydrates to form carbonic acid as:

$$CO_2(d) + H_2O \underset{k_{-1}}{\overset{k_1}{\rightleftharpoons}} H_2CO_3 \quad (31)$$

If reaction (31) is uncatalyzed and is at a pH below 8, it is freely reversible and the forward reaction proceeds at a very slow rate.

Nevertheless, the carbonic acid dissociates in an almost instantaneous ionic reaction as:

$$H_2CO_3 \underset{k_{-2}}{\overset{k_2}{\rightleftharpoons}} HCO_3^- + H^+ \quad (32)$$

Addition of reactions (31) and (32) thus yields the overall reaction:

$$CO_2(d) + H_2O \underset{k_{-3}}{\overset{k_3}{\rightleftharpoons}} HCO_3^- + H^+ \quad (33)$$

The overall equilibrium constant becomes

$$K = K_3/K_{-3} = [H^+][HCO_3^-]/[CO_2] \quad (34)$$

Reaction (33) is forced to the right as a result of the constant diffusion of CO_2 into the external environment of the cell. The bicarbonate that has been produced also undergoes one more reaction in solution to form the carbonate ions and hydrogen ions.

$$HCO_3^- \underset{k_{-4}}{\overset{k_4}{\rightleftharpoons}} H^+ + CO_3^= \quad (35)$$

Reaction (35) is extremely slow and is sometimes considered negligible. Rate constants have been determined for these reactions and are given below [22–26].

$$K_2/K_{-2} = 1 \times 10^{-3} \quad (36)$$

$$K_3/K_{-3} = 2.5 \times 10^{-4} \quad (37)$$

$$K_4/K_{-4} = 6 \times 10^{-11} \quad (38)$$

The total amount of carbonic species present in the system is represented by C_T.

$$C_T = [CO_2] + [H_2CO_3] + [HCO_3^-] + [CO_3^=] \quad (39)$$

Note that there are five unknowns in this equation, three of which are independent variables and two being the measurable quantities, i.e., C_T and the pH of the aqueous solution.

It has been found that dissolution of carbon dioxide is not pH-dependent, however the bicarbonate concentration increases with an increased pH. In addition, for pH at 5.5 to approximately 6.5 the carbonate ions are practically non-existent.

Chemical reactions have been known to affect transport of reacting species. Facilitated transport is that transport mechanism in which the flux of the diffusing species is in the direction of decreasing concentration, but at a much higher value than that can be accounted for by the passive diffusion alone. Otto and Quinn examined the facilitation of carbon dioxide through bicarbonate solutions [27]. The hydrated CO_2 in the form of bicarbonate ions (HCO_3^-) diffuses along with the dissolved CO_2. These researchers employed immobilized liquid films having a thickness of about 100 microns for their study. In addition, an enzyme that serves to enhance the transport of carbon dioxide called carbonic anhydrase was also investigated by the experimenters. Seven distinct chemical species were present in all media; they are dissolved carbon dioxide, carbonic acid, bicarbonate ions, carbonate ions, hydrogen ions, hydroxyl ions, and alkali metal ions.

Among the conclusions founded, one indicated that the metal ion concentration affected the facilitation of the dissolved CO_2. Apparently, the metal ion concentration fixes the average pH of the solution which in turn determines the ratio of bicarbonate ions to the dissolved CO_2. It was also found that the greater the amount of bicarbonate ions, the greater the extent of facilitation. Finally, the thickness of the film affected the extent of the reaction. This finding was explained by these researchers in terms of residence times. If the liquid film is thick, it takes the reactive species a longer time to interact; thus, influencing the ultimate progress of the reaction [27].

The catalyzed reactions, those reactions involving carbonic anhydrase, were studied at enzyme concentrations extending far beyond measurements recorded by other efforts [28, 29]. At small initial concentrations of carbonic anhydrase, the further addition of the enzyme increases the fractional approach to equilibrium. When the enzyme concentrations were greater than 0.5 mg ml^{-1}, further enzyme addition served to decrease the fractional approach to equilibrium and then assumed a constant value independent of the carbonic anhydrase concentration (Fig. 7).

Tsao studied the effect of carbonic anhydrase on the absorption of carbon dioxide [30]. The work was later challenged by Alper et al. [31, 32] and Donaldson et al. [33]. These papers provide a comprehensive treatment of studies with the carbonic anhydrase and utilizes many different experimental systems such as the wetted wall, stirred tank, etc.

Buffers are often used in bioreactors to prevent drastic or otherwise intolerable pH shifts in the medium. Meldon et al. [34] demonstrated that the buffering capacity of

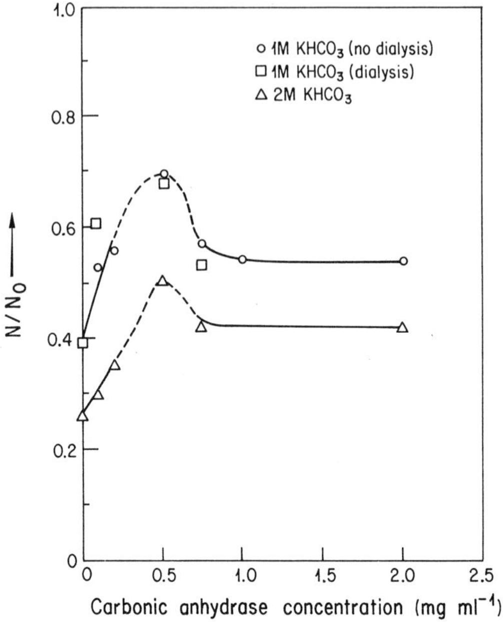

Fig. 7. Ratio of measured flux (N) to the equilibrium flux (N_0) as a function of carbonic anhydrase concentration, showing nonlinear dependence of enzyme activity on enzyme concentration. Data shown are for a liquid film thickness of 630 μm

weak acids having a pKa comparable to the solution pH can substantially enhance the transport of carbon dioxide through alkaline liquid films. The partially dissociated acids are able to couple to CO_2 hydrolysis reactions by influencing the local pH value at any point in the film.

High concentrations of carbon dioxide tend to lower the pH near a liquid interface relative to regions of lower CO_2 concentrations. The addition of a buffer, such as a weak acid or base, tends to shift the solution equilibrium to reduce the pH gradient. Lander et al. [35] carried out experiments which measured the flux of carbon dioxide across a thin liquid layer of bicarbonate solution. The pH gradient that retarded the CO_2 flux was reduced by the addition of a weak acid buffer. As a result, the rate of CO_2 transport could be increased to such a degree that a uniform pH profile was approached. Matson et al. [36] also found that the equilibrium between bicarbonate ions and carbonate ions serves to buffer the transport of weak acids such as gaseous hydrogen sulfide, H_2S, in an immobilized film.

Carbon dioxide may dissolve into caustic solutions such as sodium hydroxide, potassium hydroxide, and lithium hydroxide. The reactions involving CO_2 are described by (40) and (41).

$$CO_2 + OH^- \rightleftharpoons HCO_3^- \tag{40}$$

$$HCO_3^- + OH^- \rightleftharpoons CO_3^= \tag{41}$$

Carbon dioxide may be chemically bound to the various proteins present in a complex medium. For example, carbamates can arise by the interaction of dissolved CO_2 and the free amino groups of the proteins.

$$\begin{array}{c}R_1\\ |\\ R-C-COOH\\ |\\ NH_2\end{array} + CO_2 \rightleftharpoons \begin{array}{c}R_2\\ |\\ R-C-COOH\\ |\\ H-N-COOH\end{array} \qquad (42)$$

This reaction is favoured at pH values above the isoelectric point of a protein and competes successfully with the CO_2 hydration reaction [37]. Transient exposure of a cell to dissolved carbon dioxide results in an alteration of the protein nature. When all available $R-NH_2$ groups have been saturated, equilibrium is restored but the effects from the altered proteins remain.

Carbonic acid also associates strongly with positively charged groups of proteins by a dipole-protein interaction [37]. This binding may affect the equilibrium relations of the CO_2 by reducing the carbonic acid concentration, but eventually the process does return to an equlibrium state [37].

Carbon dioxide reacts with many other elements and molecules. The few reactions presented here are typical of those reactions that tend to occur in media. Clearly, such chemical interactions can alter the equilibrium of a system as well as the overall composition of the solution. Although a complete understanding of every reaction that occurs in the complex system is desirable, it is usually not a realistic expectation at the present time. However, consideration must be given to the dominating reactions in a system if proper monitoring of a particular component must be performed. Only research and time will offer hope of discovering and distinguishing the so-called dominant reactions from those reactions that have negligible effects [70].

4 Desorptions of Carbon Dioxide in Agitated Liquids

All cells excrete waste products. Carbon dioxide is one of these that is regularly ejected by cells. Accumulations of this nature may prove to be detrimental to the cell causing product or enzyme inhibition and ultimately culminating in the death or lysis of the cell. As such, the ventilation of carbon dioxide from the culture medium is a concern in submerged bioprocesses.

Various researchers have studied the ventilation or desorption of carbon dioxide from submerged cultures. Various findings were based on the assumption that the dissolved carbon dioxide in the culture vessel was in equilibrium with the partial pressure of CO_2 in the gas stream leaving the bioreactor. It is crucial to understand what is occurring in the medium immediately adjacent to the cell. As mentioned previously, the most abundant extracellular product is carbon dioxide. In practice, carbon dioxide continuously diffuses into the extracellular space from the respiring cells present in the vessel. The next consideration is devoted to the reactions that take place between carbon dioxide and the aqueous medium. Although microbial media are very complex in nature, the most abundant component is water. The reactions considered in Sect. 3 apply for a CO_2/H_2O system and can serve to approximate conditions in an actual medium.

Yagi and Yoshida have studied the behavior and removal of carbon dioxide from culture media [38]. They assume a distribution of bicarbonate ions throughout the vessel, which in turn varies with the rate of the hydration reactions. Two limiting cases resulted: one involving a fast hydration where the bicarbonate ion was assumed to be in equilibrium with the dissolved carbon dioxide concentration any point in the liquid film and bulk liquid, and the other case concerning a slow hydration reaction where the bicarbonate ion concentration in the film is assumed to be uniform and in equilibrium with the CO_2 concentration in the bulk liquid. There were two assumptions being made in their work. One was perfect mixing in the gas phase and the other involved plug flow in the gas phase. The reasoning for the two cases to be considered was the mean driving potential for CO_2 desorption should vary with the degree of mixing in the gas phase. Calculations for the K_La depended on which assumption was considered.

The effect of pH on the K_La values was investigated. While maintaining the same operating conditions, data from other aqueous media each having different pH values was compared to the experimental data obtained from the plug flow assumption. No trend of pH dependence of the K_La for carbon dioxide desorption was found. If the hydration reaction was sufficiently rapid, the rate of CO_2 desorption should depend on the bicarbonate concentration and consequently the pH. Because no trend was obtained, it was reasoned that chemical species that would normally facilitate CO_2 transport such as carbonic anhydrase were not present.

In consequence, Yagi and Yoshida concluded that carbon dioxide desorption took place almost as a purely physical process. They also hypothesized, based on measurements with the silicone tubing method, that the partial pressure in the exit gas is in equilibrium with the dissolved CO_2 in large industrial fermentors. This was thought to have a great significance for the possible control of submerged cultures [45].

Ishizaki, Yoshida, Shibai, Hirose, and Shiro investigated this problem [22,39,40]. Their work is divided into three papers with each successive paper becoming more complex in nature.

The first paper considers a model system where CO_2 desorption is studied in the abscence of microorganisms. The researchers were careful to employ constituents of the fermentation medium that would not produce or expel carbon dioxide. Examples of such chemical species are phosphate buffers, and the starch hydrolyzate. The actual dissolution of CO_2 was expressed by the Henderson-Hasselbach equation. In addition, gaseous carbon dioxide was assumed to dissolve in the medium according to Henry's law. The results showed that the rules concerning the dissolution and dissociation of CO_2 held for the model system.

Part two added another complexity. The study was performed in two culture systems. The inosine and the glutamic acid processes were utilized. In order to calculate the dissolved carbon dioxide concentration in the liquid, the Warburg manometry method was employed. Immediately after a sampling operation, the living cells were removed. This was considered to be a satisfactory method to fix the equilibrium state of the dissociation of carbonic acid to the sample solution. Filtration was carried out using a millipore filter and a micro syringe filter holder. Data obtained using the indirect manometry method indicated that the Bunsen absorption coefficient for CO_2 decreased as the bioreaction proceeded. The filtration method had the advantage of determining the bicarbonate ion concentration, but it suffered from one drawback —

the filtered sample had no cells. To confirm that bicarbonate levels were the same in the filtered sample as in the culture medium, a direct manometry method was instituted. A culture sample was placed in a Warburg flask and the respiration of the cells was allowed to continue. When the temperature was in equilibrium, bicarbonate ions were exhausted by the addition of sulfuric acid solution. According to the results shown in Table 4, there was a close agreement between the results from the manometry method and the indirect method. Experimental results also indicated that the Henderson-Hasselbach law could not be maintained in a culture system.

The final section made use of a modified Severinghaus CO_2 electrode to investigate the relationship between the partial pressure of carbon dioxide in the effluent gas and the dissolved carbon dioxide in the medium. The dissolved CO_2 tension was recorded by this modified electrode and the effluent CO_2 partial pressure was measured continuously with an infra-red gas analyzer. The Warburg direct method was employed to determine the bicarbonate ion concentration in the culture medium. According to their results, an equilibrium relation between the effluent CO_2 partial pressure and the dissolved CO_2 tension in the liquid phase was established. A word of caution, however, should be given that the effluent CO_2 partial pressure presented in this paper was an average value of those measured at the maximum dissolved CO_2 tension at the lowest pH and at the minimum dissolved tension at the highest pH while continuous pH control was on [40]. The equilibrium relation on such a ground, as suggested by these authors, could be debatable [70].

Nyiri and Lengyel investigated the effects of temperature, agitation, aeration, viscosity, and the presence of protein and buffers on the dissolution, hydration, and chemical reactions of carbon dioxide [41]. The method used for determination of the rate of hydration was to employ a glass electrode which could monitor the activity of hydrogen ions. It was assumed that the rate of hydration of H^+ ions is equivalent to the rate of formation of HCO_3^- ions. In addition, Reaction (32) is assumed to be practically instantaneous; hence, the rate of hydration of CO_2 could be obtained.

Table 4. Warburg manometry for filtered spent medium and the Warburg direct method [39]

Experimental No.	Method	pCO_2 atm	pH	$[HCO_3^-]$ mmole liter^{-1}
I	Direct	0.100	6.75	16.10
	filtered	0.100	6.75	15.40
II	Direct	0.120	6.50	10.70
	filtered	0.120	6.50	10.70
III	Direct	0.070	6.50	3.80
	filtered	0.070	6.50	3.80

Medium of inosine production was used

It was concluded that the rate of CO_2 hydration increased with increasing temperatures from 0 to 40 °C. Similar results were obtained when the agitation was increased. It was suggested that the dissolution of carbon dioxide was enhanced by a reduction of the gas-liquid film resistance. The bicarbonate ion was influenced by the buffer components, the buffer capacity, and the viscosity of the culture medium. Table 5

Table 5. Effect of HPO^{4-} ion concentration and buffer capacity on hydration of CO_2 in aqueous solution[a]

HPO_4^- conc. molar	Buffer capacity[b]	k_{CO_2}	State of equilibrium, pH
0.022	42.1	5.0×10^{-2}	5.37
0.044	99.0	5.9×10^{-2}	5.75
0.22	600.0	7.9×10^{-2}	6.76

[a] Temperature, 25 °C; impeller tip velocity, 2400 rpm;
[b] Determined after titration with HCl of different normalities. The buffer capacity is expressed as ml of HCl necessary to produce one unit of pH change in the pH 8.0–7.0 range

shows the effect of a phosphate buffer on CO_2 retention and is expressed by its buffer capacity. Increasing the viscosity reduced the rate of hydration of the carbon dioxide. The results were in agreement with those of Aiba et al. [42].

Although it was not clearly proven, Nyiri and Lengyel [41] stated that the rate of hydration of CO_2 into the bulk liquid is independent of the partial pressure of carbon dioxide on the surface of the liquid but did depend on the CO_2 partial pressure in the gas bubbles within the liquid. No assumptions were made about the dissolved CO_2 being in equilibrium with the CO_2 partial pressure in the effluent gas.

Esener, Kossen, and Roels [43] also suspected that the dissolution of carbon dioxide in culture media was of a much higher value than was previously expected while using equilibrium relationships. When a total carbon balance was performed over a batch process, they observed that some carbon was missing and attributed this to a supersaturation of carbon dioxide or bicarbonate ions in the culture media. They also referred to similar observations by Barford and Hall [44].

In light of these findings, more work was required to determine if equilibrium exists in submerged culture systems. Smith and Ho [45] carried out a study that clearly dem-

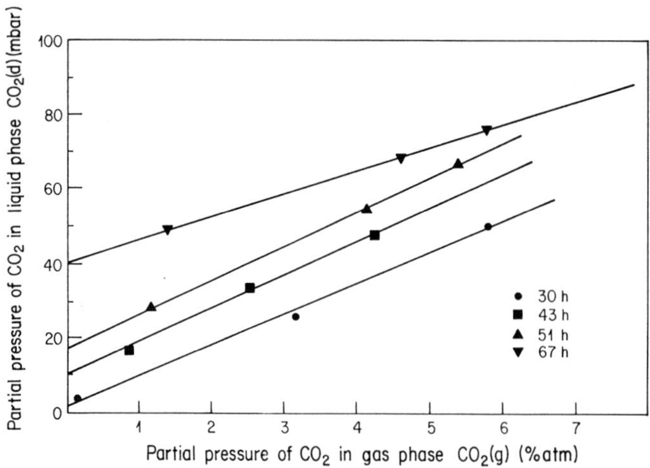

Fig. 8. Partial pressure of CO_2 in the gas phase vs. the partial pressure of CO_2 in the liquid phase

onstrates no such equilibrium exists when the culture reaches its maximum growth phase. Accordingly, any efforts to utilize an equilibrium expression given effluent gaseous CO_2 partial pressures to obtain the dissolved carbon dioxide concentration are not advisable. Instead, use of a steam-sterilizable dissolved CO_2 electrode can serve as a more accurate method in determining the accumulation of waste CO_2.

The microorganism used in their study was *Penicillium chrysogenum*. The offgas CO_2 was monitored by an infrared gas analyzer and the dissolved CO_2 was measured with an Ingold steam-sterilizable CO_2 electrode. By using the steady state gas injection method, the effluent carbon dioxide was plotted against the dissolved carbon dioxide in the media at various culture ages of the penicillin production. As demonstrated in Fig. 8, the slope remained constant at 0, 30, 43, and 51 h of the cultivation. It dropped approximately 25% when the 67 h mark was attained.

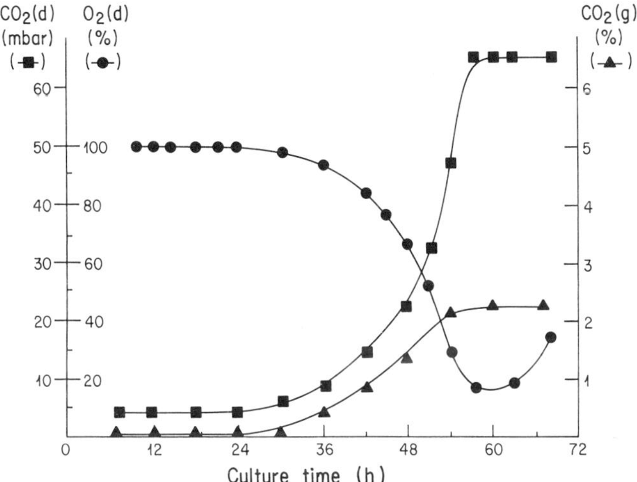

Fig. 9. Typical time profile of the disolved CO_2 and the offgas CO_2 during a penicillin production

Figure 9 shows a typical time profile of the dissolved CO_2 and the offgas CO_2 during the course of the penicillin production. In the early stages of the cultivation, values of the dissolved CO_2 and the offgas CO_2 were rather low and constant. As the culture reached its exponential growth phase, both the offgas and the dissolved carbon dioxide increased dramatically, indicating a direct relation to the quantity of viable cells.

In a typical penicillin process, the absorption coefficient was used instead of the Henry's law constant, to indicate the ratio of the dissolved CO_2 to the offgas CO_2. The absorption coefficient increased steadily up to a value of 4350 [mbar atm^{-1}] at 57 h of culture time. This was a 360% increase despite the fact that the Henry's law constants remained unchanged.

Such data illustrated that during the course of the cultivation, the dissolved CO_2 and the offgas CO_2 were not in equilibrium. Smith and Ho [45] offered a possible explanation regarding their results. During the cultivation, the cell mass increased;

thus, the total CO_2 due to cell respiration increased. As soon as the CO_2 was transferred to the culture medium, it incorporated with water to form carbonic acid. This permitted the hydrogenation reaction to take place yielding hydrogen ions and bicarbonate ions. The rate of desorption of the carbon dioxide from the medium was slower than the rate of absorption of CO_2 from the respiring cells to the liquid. The result was the birth of a state in which the total amount of CO_2 in the liquid phase increased to a lever higher than that predicted by Henry's law.

This conclusion may be substantiated by a quantitative calculation utilizing data from other researchers. Mou and Cooney reported that for *P. chrysogenum* culturing at a specific growth rate near $0.1\ h^{-1}$, the cell yield related well to the carbon dioxide evolution by 35.8 g cell per mole CO_2 [46]. The specific CO_2 evolution rate was calculated as:

$$q_{CO_2} = 1/35.8 \text{ g cell per mole } CO_2 = 1.229 \text{ g } CO_2 \text{ per g cell}$$

The overall evolution rate of CO_2 was thus:

$$Q_{CO_2} = \mu/Xq_{CO_2} = (0.1\ h^{-1})\ (50\ g\ L^{-1})\ (1.229 \text{ g } CO_2 \text{ per g}) = 6.145 \text{ g } CO_2 \text{ per L } h^{-1}$$

On examination of the data reported by Yagi and Yoshida, the mass transfer coefficient, K_La, had a value of approximately $50\ h^{-1}$ for a normal CO_2 desorption process [38]. The maximum driving force for CO_2 desorption would be the difference between the partial pressure of the dissolved CO_2 and that which was in equilibrium with the offgas CO_2. Based on the data of Fig. 9, the driving force was 0.051 atm. The equilibrium constant H of CO_2 in aqueous solution at 25 °C was approximately 1.449 g CO_2 per L-atm [50].

$$C = (0.051 \text{ atm})\ (1.449 \text{ g } CO_2 \text{ per L}) = 74 \text{ mg } CO_2 \text{ per L}$$

Therefore, the desorption rate of carbon dioxide was:

$$Q_{desorption} = K_La\ \Delta C = 3.70 \text{ g } CO_2 \text{ per L } h^{-1}$$

It is obvious that the evolution rate exceeded the CO_2 desorption rate. Therefore, during the course of the cultivation as the viable cell mass increased, excessive dissolution of carbon dioxide took place resulting in an increased absorption coefficient [15, 70].

The operation of waste removal can be a critical step in the feasibility of a submerged process. Should the scale up proceedures rely on measurements of waste gases such as carbon dioxide, studies must be conducted with care. Assumptions must be verified by experimentation and interactions occurring between the species being monitored and the complex medium should be carefully researched. The purpose of this seemingly painstaking work becomes obvious when one realizes that the product yield of a scaled up fermentation can not be dependent on false information or compounding errors.

It is equally true that such research can lead to advantages as to largely enhance product yields and cell growth by a simple manipulation of the culture environment. Still, much research remains to be completed in this area.

5 The Effect of Carbon Dioxide on Various Microbial Processes

Since it has been established that CO_2 has an effect on cellular structures and functions, it is logical to expect that the cultivations would also be affected. The work of a number of researchers pertaining to the effects of CO_2 on microbial processes will be reviewed in this chapter.

Ten strains of *Clostridium botulinum* were examined under different environmental carbon dioxide concentrations [47]. At 101 KPa or 1 atmosphere, excessive CO_2 delayed the production of toxins compared to an atmosphere of 100% nitrogen. Higher levels of CO_2 also delayed the onset of toxin production. Pressurized carbon dioxide turned out to be lethal to the microorganisms. Under pressurized conditions, the rate of decrease of recoverable colony-forming units was found to be dependent on the amount of pressure of the CO_2 and the length of exposure. Still, even at a CO_2 pressure of 816 KPa absolute, there were still surviving spores [47]

During the aerobic batch growth of yeasts, Jones and Greenfield noticed increased lag phases with CO_2 partial pressures of 0.5 bar [37]. The lag times ranged from 1 to 7 h using glucose, fumarate or succinate as the carbon substrates. Microbial growth on fumarate or succinate was more sensitive to carbon dioxide inhibition than growth on acetate, glucose, malate, or citrate.

Yeasts grown without oxygen, that is, under the anaerobic conditions, on glucose required dissolved CO_2 in order to synthesize essential amino acids. It was found that the substrate yield coefficient could be improved by 3 to 5% when dissolved CO_2 was present in the form of bicarbonate ions.

Chen and Gutmanis also investigated the inhibition of carbon dioxide on *Saccharomyces cerevisiae* grown under substrate-limiting conditions [48]. Data showed that up to 20% of CO_2 in the effluent gas caused little inhibition under the conditions of favorable biomass production. However, significant inhibition was observed when effluent concentrations of CO_2 reached 30%. At still higher concentrations, the process "activity" was ceased.

Hirose et al. [49] noted the effects of agitation and partial pressures of CO_2 on *Brevibacterium lactofermentum* cultures. Although they concluded that the changes due to the influence of carbon dioxide concentrations were insignificant compared to the effects of oxygen manipulation, there was a small decrease in product yield as the CO_2 levels were increased. The validity of the technique with which these researchers determined their CO_2 concentrations is, however, debatable.

Efforts have been devoted to correlating various operating variables to cellular processes or functioning. For example, Cooney et al. [50] found a relationship between the microbial heat production and the rate of CO_2 evolution. Experiments were carried out using *Escherichia coli*, *Bacillus subtilis*, *Candida intermedia*, and *Aspergillus niger* on different substrates including glucose, molasses, and soy bean meal. Data was obtained for eight cultivations. Despite the scatter of the plotted points, a linear

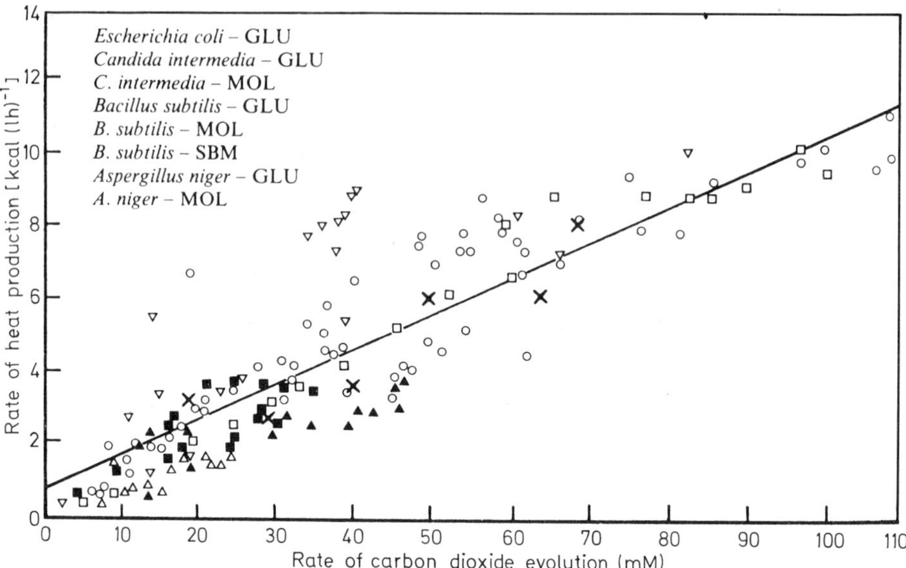

Fig. 10. Rate of heat production vs. the rate of CO_2 evolution. GLU — Glucose medium, MOL — Molasses medium, SBM — Soy bean meal medium

trend did result. It was also noted that data from a single cultivation had a tendency to group to one side of the line or the other (see Fig. 10). The variation in the ratio of the heat released to the CO_2 produced was explained by the oxidation reactions that occurred while the heat production rate and carbon dioxide evolution rate were measured.

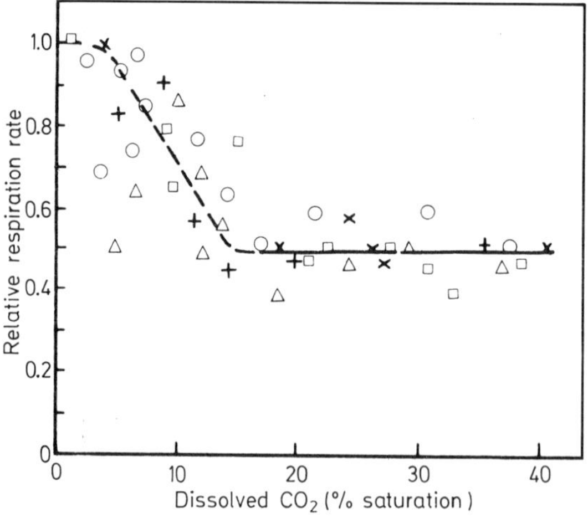

Fig. 11. Dependence of the respiration rate of the oleandomycin and tetracycline producing microorganism on the concentration of dissolved carbon dioxide. (△), (○), (+) experiments with tetracycline; (□), (×) experiments with oleandomycin

Culture media containing microorganisms producing tetracycline, oleandomycin, and streptomycin were subjected to various dissolved CO_2 levels (see Fig. 11) [51]. For both the tetracycline and the oleandomycin producers at 15% to 20% concentrations of dissolved CO_2, the respiration decreased from about 40% to 50% from the original respiratory level. Upon return to the original low CO_2 conditions, these microorganisms retrieve their normal respirations. Streptomycin producers, however, did not return to normal after they had been exposed to same dissolved CO_2 concentrations. It was thought that CO_2 caused irreversible enzymatic damage to these streptomycin-producing microorganisms.

In general, the specific rate of penicillin synthesis is of prime importance in the submerged process. To date, the optimization of the production rate remains as one of the most challenging areas in the improvement of submerged penicillin synthesis.

The inhibition of production by carbon dioxide was studied by Pirt and Mancini [52]. Using a chemostat at fixed values of pH and dissolved oxygen, they were able to control the conditions such that the only variable was the partial pressure of carbon dioxide. Pirt and Mancini found that the synthesis rate (units penicillin per mg dry mycelium per h) was decreased 50% at a carbon dioxide partial pressure of 0.08 atm compared to the control case [52].

Research regarding the effects of carbon dioxide on production rate is unfortunately scarce. In practice, it has been neglected in the biochemical literature. Ho and Smith [53] have investigated the effect of CO_2 upon submerged reactions. The following treatment will provide a comprehensive view of the effect that carbon dioxide has on an antibiotic producing microorganism, *P. chrysogenum*. The purpose of maintaining these variables constant under non-substrate limiting conditions was to ensure that carbon dioxide was the only variable. The influent gas was air with various portions of its nitrogen being replaced by equivalent amounts of carbon dioxide to constitute the influent CO_2 partial pressures of 0%, 3%, 5%, 13%, and 20%, for various cultivations.

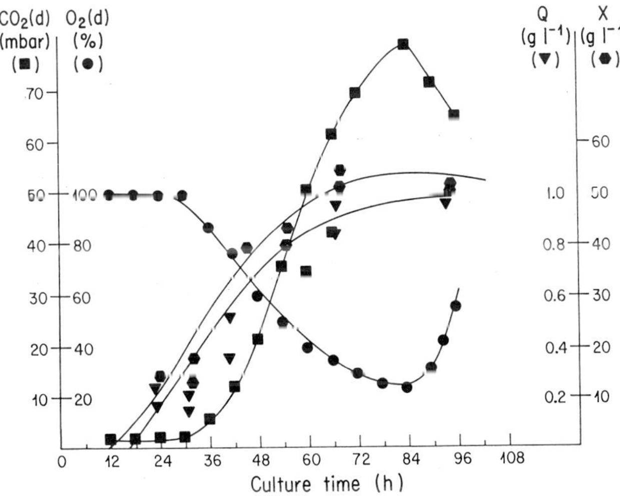

Fig. 12. Time profile of the control penicillin process

The processes subject to different partial pressures of carbon dioxide in the input gas stream were compared to the control case in which no CO_2 was added to the influent air stream [53]. Figure 12 shows the time profile for the control of the process. As can be seen, the concentration of carbon dioxide in the liquid phase was related to the concentration of viable cells in the culture medium. The peak carbon dioxide in the liquid phase occurred almost identically with the minimum dissolved oxygen level and the maximum dry cell concentration when the cultivation time was approximately 81 h. At the maximum dissolved carbon dioxide concentration of 79 mbar, the dry cell concentration was 53 g L^{-1}. The maximum penicillin concentration was found to be 1.0 g L^{-1} at 93 h.

Continuously exposing the process to an influent gas composing of 3% CO_2 resulted in little reduction in metabolic performance. Figure 13 shows the time profile of the culture sparged with the 3% CO_2 gas. Since this run was conducted at a relatively low concentration of CO_2, it was possible that the carbon dioxide partial pressure was held below a critical level for metabolic inhibition. No appreciable change was evident in either the dry cell mass or the concentration of penicillin in the culture medium.

A scanning electron microscopic analysis of the mycelial morphology indicated, however, an increased hyphal branching frequency in comparison to the hyphae of the control run [55]. An increased branching frequency was reported to be associated with high growth rates in cultures of *P. chrysogenum* grown under non-substrate limiting conditions [54]. An increase in the branching frequency could be attributed to the higher than normal concentration of carbon dioxide present in the medium during both the lag phase and the early part of the exponential growth phase. It was possible that carbon dioxide inhibition in the primary metabolism was compensated by a higher than normal branching frequency. Since *P. chrysogenum* grew and duplicated at the hyphal tips, an increase in the number of tips (or branches) would yield a higher rate of mycelial growth. Thus, the inhibition of cell growth by the increased carbon dioxide concentration was offset by an increased branching frequency.

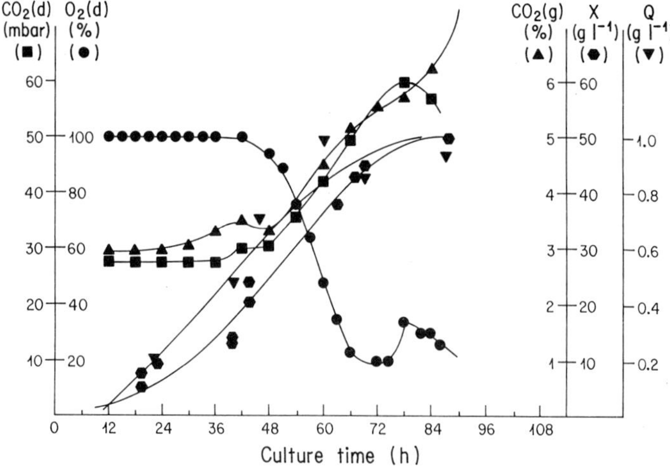

Fig. 13. Time profile of the penicillin culture sparged with the influent gas composing of 3% CO_2

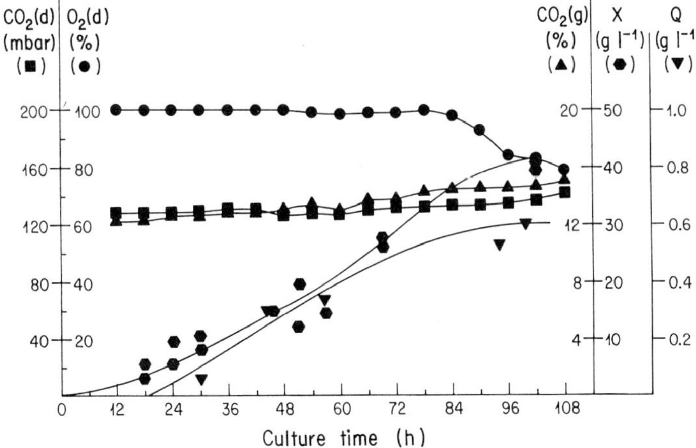

Fig. 14. Time profile of the penicillin process sparged with the influent gas composing of 13% CO_2

A dramatic decrease in biological activities occurred while the culture was injected with an influent gas composing of 13% CO_2. Figure 14 shows the time profile for *P. chrysogenum* grown under contact with the 13% CO_2 gas. As can be seen from the figure, cell growth was significantly reduced 25%, and the time to reach the maximum cell concentration was extended by 21 h in comparison with that of the control run. Likewise, the penicillin production was inhibited by a factor of 40% from the reference maximum value which was delayed for 9 h.

The occurrence of highly branched mycelia, pellets, and swollen hyphae was observed through the scanning electron microscope. The presence of pellets and swollen hyphae was a direct result of the exposure to excessive carbon dioxide present in the culture medium. This in turn yielded a profound effect on the cell metabolism and morphology. The appearance of the two morphological forms was significant not only in their occurrence, but also in the large number of pellets and swollen hyphae that were observed. These two forms have been reported to be associated with low growth rates [54)], and the presence of these morphological forms in the reactor most certainly influenced the penicillin producing capability of *P. chrysogenum*.

The inhibitory nature of carbon dioxide on cell metabolism and penicillin production was demonstrated while the penicillin cultivation was operated under the continuous contact with the 20% carbon dioxide gas. Figure 15 shows the time profile of this culture process. The maximum cell concentration was 16.5 g L^{-1} observed at 96 h of cultivation time, and the cell growth had not reached a stationary phase. The maximum penicillin concentration was only 0.1 g L^{-1} at a cultivation time of 96 h. These low levels of cell and penicillin concentrations indicated a dramatic inhibition on both the primary and secondary metabolisms as a result of exposure to the excessive carbon dioxide. The scanning electron microscopy revealed an extensive quantity of swollen and highly branched hyphal structures with almost no healthy or normal cells present. The morphological change seemed to be similar to that which occurred during the cultivation exposed to the 13% CO_2 gas with the exception that larger swollen cells were present to a much higher degree. The dominant morphological

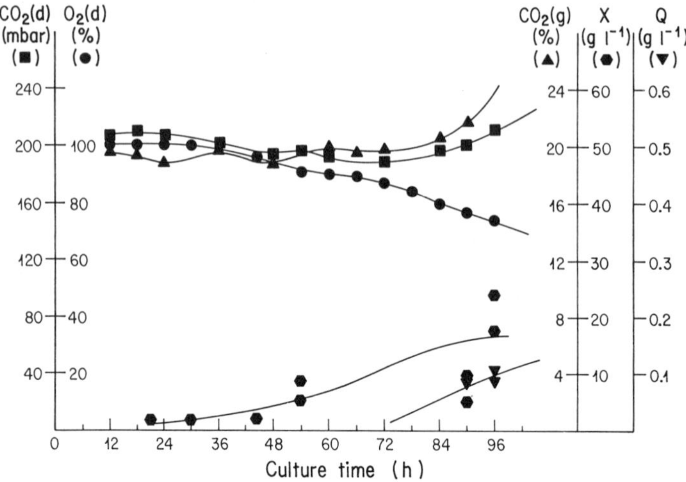

Fig. 15. Time profile of the penicillin process sparged with the influent gas composing of 20% CO_2

form was the swollen, short, thick, and highly branched hyphal structures. The presence of these altered morphological forms, marked by a low metabolic activity, clearly indicated a high degree of carbon dioxide inhibition.

For the submerged penicillin process, the concentration of the dissolved CO_2 was increased, the cell mass and penicillin production were reduced to varying degrees depending on the magnitude of carbon dioxide intoxication. In Table 6, the maximum cell mass and product concentration are summarized for various cultivations [53, 70].

For each cultivation the specific growth rate was affected by the presence of carbon dioxide in the medium. As the level of carbon dioxide was increased from 0–5%, the growth rate of *P. chrysogenum* remained constant. At carbon dioxide partial pressures of 13–20%, the specific growth rate of cells fell to a CO_2 inhibited level. Presented in Fig. 16 are the specific growth rates for the various cultivation runs. The specific growth rates at carbon dioxide partial pressures between 0–5% appeared to be fairly constant. When the penicillin culture medium was subjected to a steady state injection of the 13–20% carbon dioxide gas, the specific growth rate was dramatically reduced.

Table 6. The maximum cell and penicillin concentration and the corresponding culture times for various processes

CO_2 partial pressure	Cell concentration		Product concentration	
	Max. reached	Max. at reaction time	Max. reached	Max. at reaction time
(% atm)	($g L^{-1}$)	(h)	($g L^{-1}$)	(h)
0	55	81	1.0	90
3	50	84	1.0	81
5	50	96	0.95	93
13	41.5	102	0.60	99
20	16.5	96	0.10	96

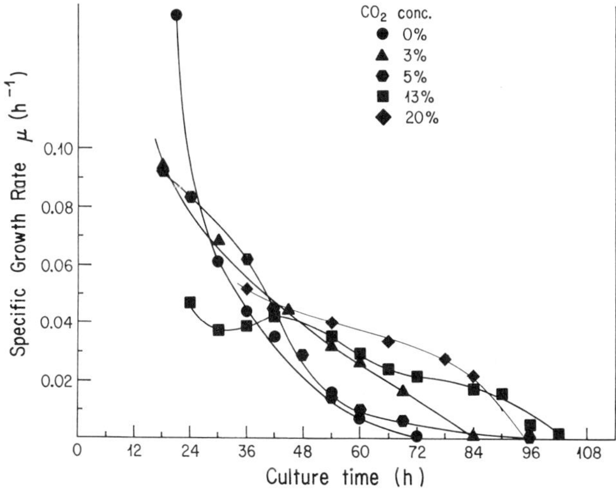

Fig. 16. Specific growth rates of *Penicillium chrysogenum* cultivated at various concentrations of carbon dioxide. (—○—, —▲—, —●—, —■—, and —◆—) Indicate respectively the carbon dioxide partial pressures of 0%, 3%, 5%, 13%, and 20% in the influent air stream

Evidently the growth rate of *P. chrysogenum* was affected by high levels of carbon dioxide present in the medium, though at low levels of carbon dioxide there appeared to be little change in the specific growth rate. At high concentrations of carbon dioxide a complete and seemingly irreversible change in the cell morphology occurred; therefore, the reduction in the primary metabolic efficiency was easily explained. As mentioned earlier, the scanning electron microscopy of the mycelial organisms cultivated at low levels of carbon dioxide, i.e. 3% and 5%, revealed a significant increase in the hyphal branching frequency. Since hyphae grew actively only at the tips, as the

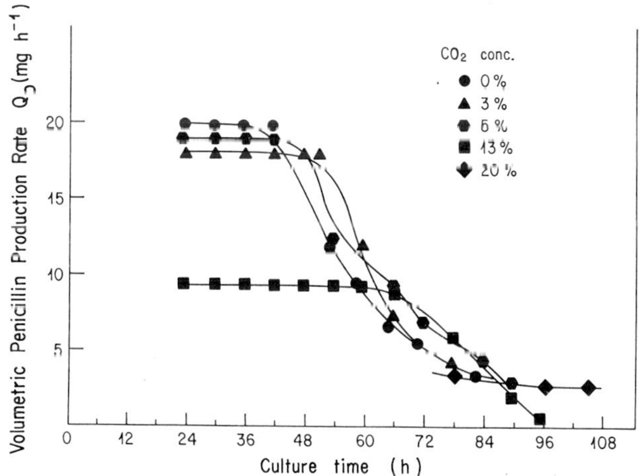

Fig. 17. Effect of the carbon dioxide concentration on the volumetric penicillin production rate

number of tips (or branches) increased, so would the rate of cell growth. Under this condition, any inhibition of cell metabolism could be easily offset by the added advantage of an increased number of growth sites. Cells grown at high concentrations of carbon dioxide appeared stunted and deformed indicating altered metabolic functions. Figure 17 shows the volumetric penicillin production rate for the various batches continuously sparged with gases of different compositions of CO_2. It could easily be seen that the high carbon dioxide partial pressures present in the medium seriously reduced the penicillin producing capability of *P. chrysogenum*. When injected with the 13% CO_2 gas, the volumetric production rate was reduced by 50% during the initial production period. Injection with the 20% carbon dioxide gas almost totally interrupted production as the potential of cells was reduced almost to nil. At lower CO_2 levels, the penicillin producing capability was little affected.

It was interesting to note that the dominant morphological form was filamentous at low concentrations of carbon dioxide while the production was close to normal. At higher levels of dissolved carbon dioxide, the presence of severely deformed and swollen hyphae associated with low rates of synthesis led to the elucidation that carbon dioxide had a threshold value for the inhibition. Similarity between this phenomenon and that observed in the specific growth rate analysis demonstrated that the dissolved carbon dioxide had profound consequences in the submerged antibiotic cell growth and product formation.

The important study pertaining to the effect of CO_2 on all cellular processes and hence on submerged aerobic processes just at its beginning. Much concerted research must be devoted to the areas of microbiology, microbial technology, and chemical engineering if a thorough understanding is to be realized [70].

6 The Effect of CO_2 on Mycelial Morphology

The morphology of submerged mycelial organisms is of prime importance in the evaluation of cell growth and product formation for important antibiotic industries. Cells which have undergone morphological alterations due to an adverse environmental condition or the presence of toxins are more likely to exhibit a decreased metabolic rate than those which are free from such morphological changes. In consequence, the importance of a morphological study can never be over-emphasized in the submerged antibiotic processes.

It is well known that morphology of submerged mycelial organisms can be greatly affected by a variety of environmental variables. For example, the effect of pH on mycelial morphology was examined by Pirt and Callow in a continuous submerged process [57]. It was determined that pellet formation and formation of the normal filamentous forms of *P. chrysogenum* could be influenced by the pH of the medium. Furthermore, the length of the hyphae varied considerably with pH. They determined that the average hyphal branch length was 200 μm for mycelia grown at pH 6.0 compared to a length of 20 μm at pH 7.4. At pH 7.4, however, they observed large numbers of swollen yeast-like cells, indicating the beginning of pellet formation. They reported that the morphological alteration would return to normal with a decrease in pH; in other words, the morphological change due to pH was reversible.

Miles and Trinci recently studied the effect of pH on the length and diameter of hyphal branch units of *P. chrysogenum* grown in glucose-limited chemostat cultures at 25 °C and a dilution rate of 0.09 h^{-1} [56]. The hyphal branch length varied with pH, attaining a maximum value of 110 μm at pH 6.0 compared to that of 45 μm at pH 8.0. The variation of the diameters of hyphal grown at different pH values was found to be insignificant. They suggested that pH 6.0 be the optimum pH for hyphal extention of *P. chrysogenum*.

Pirt and Callow also demonstrated that the pH was not the only factor that altered cell morphology [57]. Substitution of the nitrogen source of ammonium sulphate by corn steep liquor increased the hyphal length, reduced the frequency of hyphal branching, and prevented the formation of swollen cells at pH 7.4. Nutrient formulation was shown to be an important factor as well. For instance, Morton [58] observed that conidiation of *Penicillium griseofulvum* was rapidly induced in submerged conditions when the mycelia were placed in a medium containing high concentration of glucose but without assimilable nitrogen. He reported, however, that upon nitrogen exhaustion, *P. chrysogenum* remained obstinately vegetative.

The specific growth rate was also proven to affect cell morphology. Righelato et al. [54] found that vegetative growth occurred at a specific growth rate of 0.023 to 0.075 h^{-1}. They found that as the growth rate was increased within this range, the occurrence of pellets and swollen cells increased. At specific growth rates of 0.014 h^{-1} and below, conidiation occurred.

In an earlier study, Smith and Ho [55] demonstrated that upon exposure to moderate to high dissolved CO_2 concentrations the penicillin cultures revealed tremendous morphological changes under scanning electron microscopic examinations. In what follows, a brief review of the findings and electron micrographs of a wide spectrum of morphological forms of *P. chrysogenum* obtained by growing cultures at various carbon dioxide concentrations will be presented.

Figure 18 shows the lyopholized spores by *P. chrysogenum*. These were characteristic-

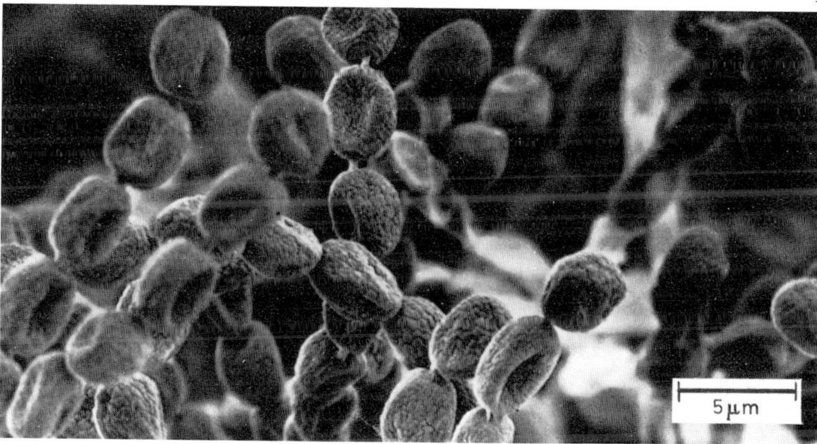

Fig. 18. SEM micrograph of lyopholized spores of *Penicillium chrysogenum*. Magnification = 4000 ×

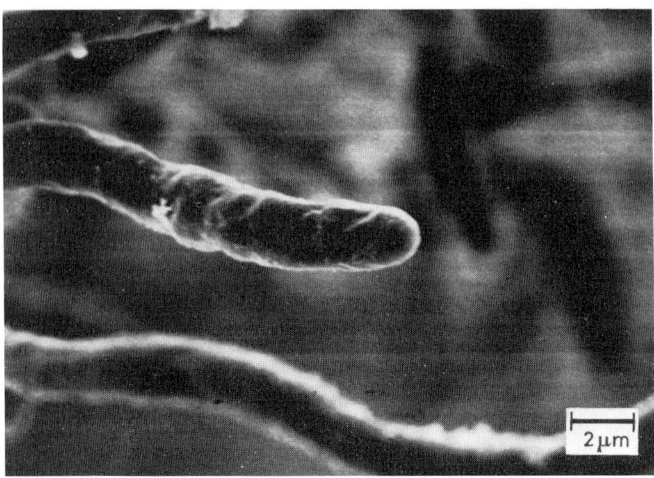

Fig. 19. SEM micrograph of the *Penicillium chrysogenum* for the control cultivation. The mycelia were harvested at 48 h of cultivation. Magnification = 5400 ×

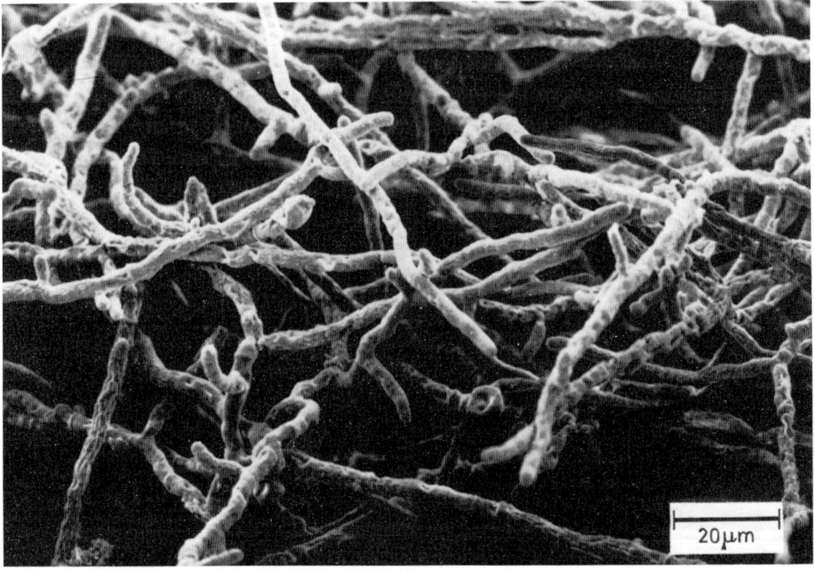

Fig. 20. SEM micrograph of the filamentous mycelial organisms obtained by exposure to the 8.2% carbon dioxide gas. The mycelia were harvested at 48 h of cultivation. Magnification = 900 ×

ally asexual conidiospores being produced by *P. chrysogenum* on agar slant cultures. The size of the spores were 5.5 μm.

Presented in Fig. 19 is an SEM micrograph of mycelial structure of a healthy *P. chrysogenum* grown under a control condition where no CO_2 was added to the influent air. The mycelia were harvested after 48 h of cultivation. As can be seen the mycelial structure was composed of a network of smooth hyphae. In this specimen, the hyphae are long, thin, and diffuse. The average thickness of the hyphae was 2.0 μm. Characteristically, the hyphae grew from a central branch, and the active growth occurred at hyphal tips (Pirt and Callow [57], Righelato et al. [54], Trinci and Righelato [59]; Collinge et al. [60], and Miles and Trinci [56]). To a great extent, the rate of mycelial growth was determined by the quantity of active hyphal tips present in the culture. In a typical experiment, the desired morphological form is of the filamentous nature as presented in Fig. 19.

Presented in Fig. 20 is an SEM micrograph of mycelia grown under influent gas having carbon dioxide partial pressure of 8.2%. The mycelia were harvested at 48 h. The hyphal branching frequency seemed to be greater than that for cells grown under the control experiment. The thickness of the hyphae was 2.0–2.5 μm. Examination of Fig. 20 showed that the morphological form was predominantly filamentous with normal linear growth occurring. Analysis of the mycelia revealed an increase in hyphal branching frequency which, to a large extent, would result in a high growth rate as the quantity of active growing sites was increased.

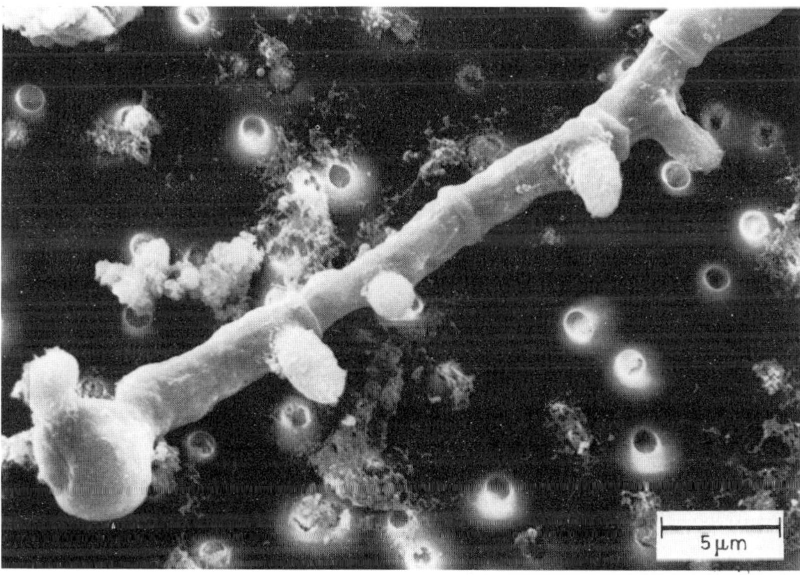

Fig. 21. SEM micrograph of the *Penicillium chrysogenum* hyphal branch exposed to the 8.2% carbon dioxide gas. Magnification = 4000 ×

The increase in the hyphal branching frequency is demonstrated in Fig. 21. The original growth initiation site, the spore, is located in the lower left-hand corner of Fig. 21. At this original site linear growth occurred. As the hypha aged, new growth sites generated along the hypha, and these sites appeared as the small buds present in Fig. 21. These buds continued to extend and would develop into new branches.

Cells grown under contact with an influent gas containing 15% carbon dioxide

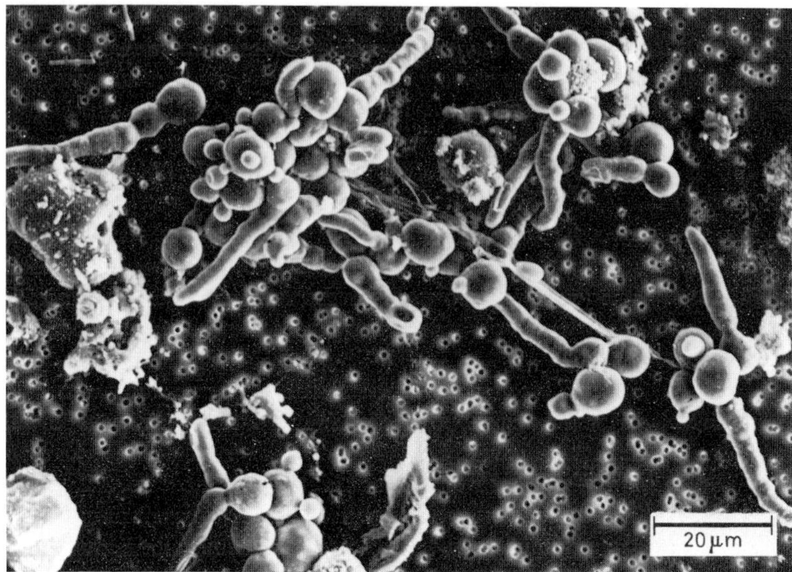

Fig. 22. SEM micrograph of the swollen spores and hyphae resulting from exposure to the 15% carbon dioxide gas. Magnification = 1000×

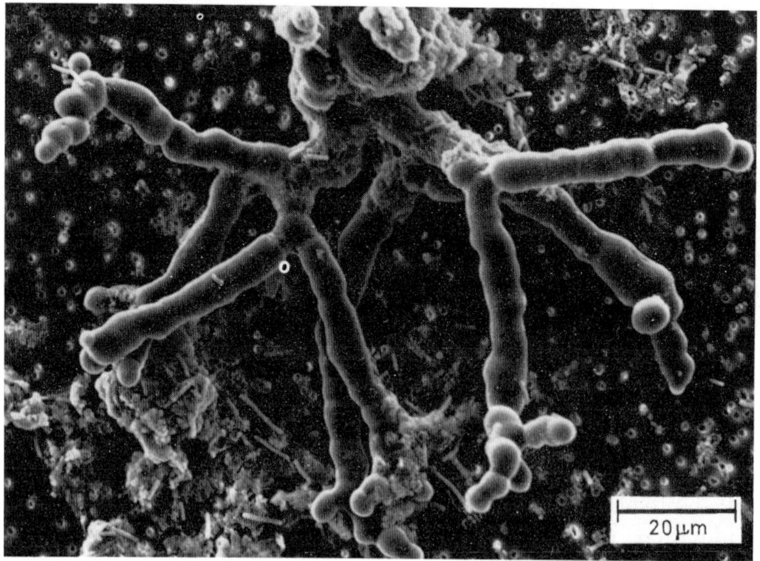

Fig. 23. SEM micrograph of the *Penicillium chrysogenum* mycelial structure resulting from exposure to the 20% carbon dioxide gas. The mycelia were harvested at 56 h of cultivation. Magnification = 1000×

were found to have morphology similar to yeast cells and spores. Presented in Fig. 22 is an SEM micrograph of mycelial cells grown under contact with the 15% carbon dioxide gas. These mycelial cells were harvested at a culture age of 48 h. The mycelial morphology was altered to what appeared to be spherical growth or a form of yeast-like morphology. Some filamentous growth was noted. The linear growth originated from the spores which looked similar to the spherical or yeast-like forms present in the culture medium. It was possible that the spherical forms were simply stunted spores. Given sufficient time these spores would develop into hyphae with a significantly altered morphological form. The thickness of the hyphae was 2.5–3.0 μm. The origin of hyphal growth (the spore) was approximately 6.0–7.0 μm in diameter. The thickness of the hyphae and the spore represented an increase of 20% and 12%, respectively, compared to the control cultures.

Cells grown at an influent gas containing 20% carbon dioxide are shown in Fig. 23. The mycelia were harvested at an age of 56 h. It can be seen from the micrograph that a significant change occurred in the cell morphology. At this influent carbon dioxide partial pressure, the individual cells within the hyphae were swollen to a degree that they were clearly detectable by direct examination of the hyphal cell wall. This was characterized by the bumpy surface on the cell wall. The thickness of the hyphae was between 4.0 and 7.0 μm. The thickness of the orginal growth site (the spore) could not be measured because of difficulties in identification.

Shown in Fig. 24 are hyphae grown under contact with an influent carbon dioxide partial pressure of 20%. The mycelia were harvested at a culture age of 88 h. Under the exposure of high CO_2 concentrations, the *P. chrysogenum* cultures consisted of swollen hyphae with which individual cells were identifiable by the bumpy surface on

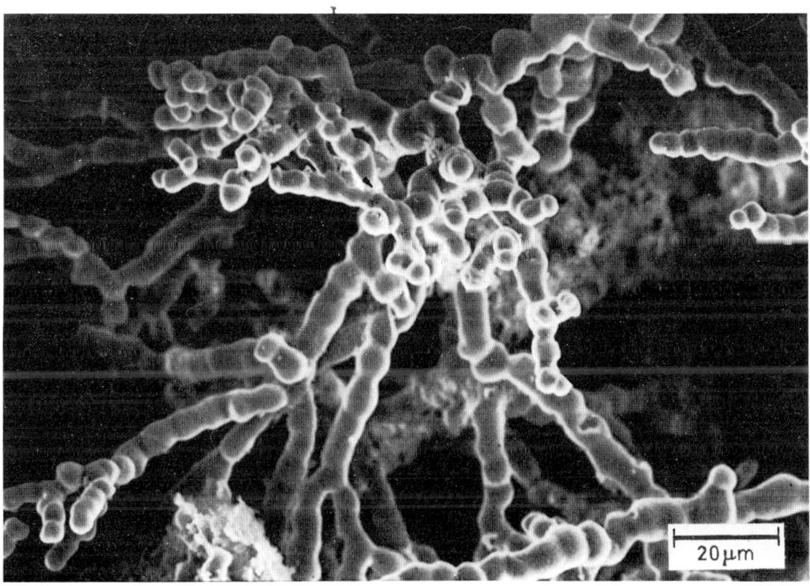

Fig. 24. SEM micrograph of the swollen hyphae grown under exposure to the 20% carbon dioxide gas. The mycelia were harvested at 88 h of cultivation. Magnification = 900×

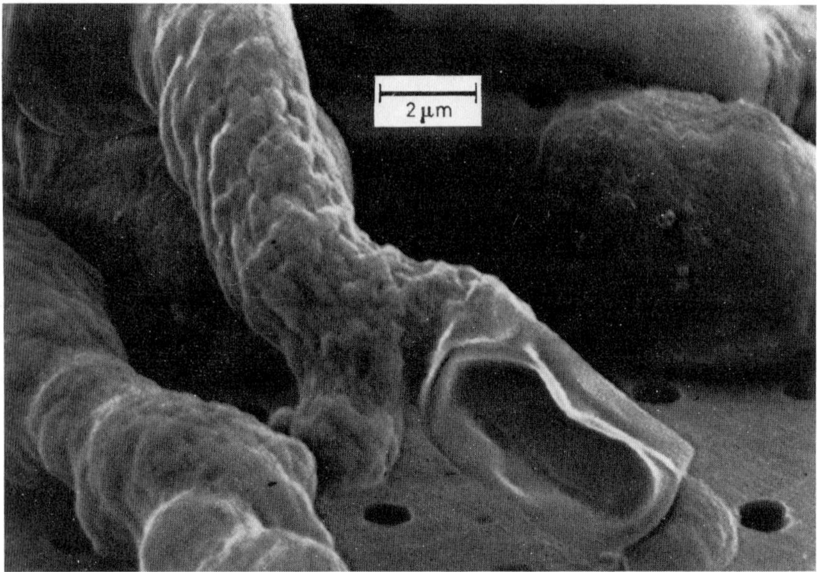

Fig. 25. SEM of *Penicillium chrysogenum* exposed to 20% carbon dioxide. Note the punctuated cell wall. Harvest time was 88 h of process time. Magnification = 8200 ×

the cell walls, as shown in Figs. 24 and 25. This suggested that conidiation occurred in the presence of excessive carbon dioxide. In typical submerged mycelial cultures conidiation would normally occur under adverse environmental conditions (Morton [58]; Righelato et al. [54]; Bent and Morton [61]). The mycelial organisms were of the aberrant form as described by Pirt and Callow [57], Morton [58], and Bent and Morton [61]. However, it should be noted in Fig. 25 that the mycelial organisms not only had their hyphae tremendously deformed, but also had their cell wall punctured. The situation was in contrast with the cultures harvested at 56 h as there were no holes being detected on the cell wall of the younger cultures. This indicated that the length of exposure to excessive CO_2 could be responsible for the defective cell wall as observed in the cultures.

Upon close examination of the different morphological forms obtained by exposing the cultures to various carbon dioxide concentrations several comments could be made. First, as the concentration of carbon dioxide in the medium increased, so did the degree of hyphal swelling. This indicates that either the excessive carbon dioxide affected the membrane transport properties of the cell or the mycelial cells were subject to osmotic swelling. It was reported that carbon dioxide could alter the cell membrane by affecting the fatty acid core. Likewise, the bicarbonate ion could affect the charged surface of the phospholipid head groups and proteins at the surface of the membrane (Albert et al.[62]; Houslay and Stanley [63]). Therefore, it seemed likely that the dissolved carbon dioxide present in the medium was responsible for hyphal swelling as a result of the alteration of the membrane transport processes. Continued accumulation of metabolites or water might result in further swelling.

It was known that many fungi have the ability to grow in either the filamentous form or in a yeast-like form depending on environmental conditions. The property is referred to as fungal dimorphism (Stewart and Rogers) [64]. A major result of the present research was that carbon dioxide was demonstrated to be a prime factor causing dimorphism in *P. chrysogenum*. The mycelial organisms could develop into a yeast-like form in the presence of excessive carbon dioxide. Under the exposure of low to moderate levels of carbon dioxide, the cultures caused predominantly filamentous growth to occur with some yeast-like cells present depending on the concentration of carbon dioxide and the culture age. The occurrence of the yeast-like cells would be due to the hyphal swelling as a result from the alteration of cell membrane caused by the presence of excessive carbon dioxide.

Another interesting phenomenon worth hypothesizing might occur at the chitin synthesis sites along the cell wall. The chitin synthesis and the resulting chitin fibrils largely determined the shape or direction of the mycelial growth (Stewart and Rogers) [64]. Linear growth would occur as the chitin fibrils formed a microfibril net around the circumference of the hyphae. The pressure exerted by the growing cells resulted in a condition where the hyphae must expand in a linear direction. Spherical growth would result from the elimination of rearrangement of the chitin synthesis sites, and ballooning would occur as the cell internal pressure increased. It was quite possible that as the concentration of carbon dioxide was increased in the medium, the chitin synthesis sites were reduced and the spherical or yeast-like growth occurred. At low carbon dioxide concentrations, the chitin synthesis sites would be active and largely unaffected by carbon dioxide leading to a normal filamentous growth pattern.

The study by Smith and Ho [55] revealed that under exposure to low carbon dioxide influent gases the branching frequency of growing hyphae increased, as demonstrated in Fig. 21. The increase in hyphal branches might somehow be related to a higher trans-cellular electric current (Jaffe [65]; Harold et al. [66]) the *P. chrysogenum* cultures experienced when sparged with low CO_2 gases. Jaffe [65] and Harold et al. [66] reported that many kinds of tip-growing fungi generated endogenous electric currents which could play important roles in orchestrating fungal growth and differentiation. Jaffe and Muccitelli [67] were able to demonstrate by using a tiny vibrating probe that a variety of fungi drove electric currents through themselves. They observed that a current of protons entered the apical zone of growing hyphal tips and left distally. Careful measurements with pH microelectrodes made by Kropf et al. [68] and Kropf et al. [69] also revealed that the medium immediately adjacent to a growing tip was slightly more alkaline than the bulk phase. These researchers attributed this to a flow of protons into the growing tip. Jaffe [65], Harold et al. [66], and Kropf et al. [68, 69] conclusively stated that an inward current of protons often preceded the outgrowth of new tips and could accurately predict the site of emergence. In consequence, it would be qualitatively true that a higher branching frequency could be anticipated provided more protons be available for generation of endogenous electric currents for the growing hyphae.

When a low CO_2 containing gas was sparged to sumberged *P. chrysogenum* cultures, it incorporated with water to form carbonic acid and was readily coverted to H^+ and HCO_3^-. Under such circumstances, a higher quantity of protons was present in the medium immediately surrounding the growing hyphae, thus leading to a higher hyphal branching frequency as compared to that of the control cultivation.

6.1 Conclusions for Applications

A biological specimen preparation method for scanning electron microscopic (SEM) studies worked very well as far as the fixation and dehydration of the mycelia were concerned. The SEM study further revealed and confirmed that the mycelial morphology was subject to change while *P. chrysogenum* was sparged with gases of various CO_2 partial pressures. At low partial pressures of carbon dioxide, 3–8 % in the influent gas stream, an increase in the hyphal branching frequency was qualitatively established. However, at these low concentrations of carbon dioxide the morphological form was filamentous in nature. A significant morphological change occurred when cells were grown at carbon dioxide partial pressures of 15 % and 20 %. At these high CO_2 contents swollen and stunted hyphae predominated, and the appearance of spherical or yeast-like cells was observed. Conidiation occurred under such adverse conditions.

Further research should be directed toward a fundamental elucidation on a cellular basis of the alteration of mycelial morphology caused by the presence of carbon dioxide. A better understanding of the morphological phenomenon can make possible a more rational approach to the improvement of the industrially important antibiotic processes.

7 Conclusions

A better understanding of carbon dioxide transfer in biochemical reactors is of critical importance and would make possible a more rational design and scale-up of industrially importance submerged biotechnological processes. Unfortunately, the research in this subject area has drawn little attention in the past, and much more fundamental information is clearly required.

In such a brand new field of research, it is obvious that a broad spectrum of subjects will fascinate researchers having different scientific backgrounds. Progress in research into the feature subject will depend, to a great extent, on the continuation and extension of investigations of the types reviewed in this article. It would only be fair to state that much more work remains to be done experimentally and theoretically in the CO_2-related research in biochemical engineering. This is not a simple problem, and it certainly will tax the ingenuity of biochemical researchers for some time.

8 Acknowledgement

Thanks are due to Dr. Fredric G. Bader of the Bristol-Myers Co. for valuable discussions and comments.

9 Nomenclature

A reactor cross section perpendicular to the gas flow (cm^2)
a interfacial area per unit volume ($cm^2\ cm^{-3}$)
C constant (dimensionless)

C_a concentration of component a (mg L^{-1})
C_a^* concentration of component a at gas-liquid interface (mg L^{-1})
C_g concentration of a component in gas phase (mg L^{-1})
C_n concentration of organic solute (g L^{-1})
C_{el} concentration of electrolyte in solution (mol L^{-1})
C concentration of dissolved gas in bulk liquid (g cm^{-3})
C* interfacial dissolved gas concentration in equilibrium with the bulk gas (mol cm^{-3})
C_g bulk gas concentration (mol cm^{-3})
C_g^* interfacial gas concentration in equilibrium with the liquid concentration (mol cm^{-3})
c_i concentration of ionic species i in solution
c_{O_2} average concentration of dissolved oxygen (mol cm^{-3})
D_{AB} diffusivity of A in B (cm^2 s^{-1})
D_A diffusivity of solute (cm^2 s^{-1})
D_{O2} oxygen diffusivity in liquid phase (cm^2 s^{-1})
d turbine diameter (cm)
F_g gas flow rate (cm^3 s^{-1})
H Henry's Law Constant
H_i parameter (L mol^{-1})
h empirical parameter (L mol^{-1})
h_- empirical parameter (L mol^{-1})
h_+ empirical parameter (L mol^{-1})
I_i ionic strength (mol L^{-1})
K constant (dimensionless)
K solubility parameter for non-electrolytes (L g^{-1})
K Sechenov constant
K_L overall mass transfer coefficient (cm s^{-1})
k_g local gas side mass transfer coefficient (cm s^{-1})
k_l local liquid side mass transfer coefficient (cm s^{-1})
M_B molecular weight of solvent (g mol^{-1})
m parameter (dimensionless)
N rotational speed of turbine (revolutions per min)
n parameter (dimensionless)
P_g gassed power input (g cm s^{-1})
p_a partial pressure of component a (KPa)
p_{ai} partial pressure of component a at the interface (KPa)
Q_i local uptake rate (mole cm^{-3} s^{-1})
$\bar{Q}$ average uptake rate (mole cm^{-3} s^{-1})
R rate of absorption (mole cm^{-2} s^{-1})
$\bar{R}$ average rate of absorption (mole cm^{-2} s^{-1})
R′ specific CO_2 production rate (mol CO_2 per mg cell per s)
—r transfer rate of CO_2 across gas-liquid interface (mol CO_2 per cm^3 s^{-1})
s surface renewal rate (s^{-1})
T temperature (K)
V reactor volume (cm^3)
V_A molecular volume of solute (cm^3 mol^{-1})

V_L liquid volume (cm^3) or liquid flow rate ($cm\ s^{-1}$)
V_S superficial gas velocity ($cm\ min^{-1}$)
V_T terminal bubble rise velocity ($cm\ min^{-1}$)
X' viable cell mass (mg cell per cm^3)
z_i valencies of ions (dimensionless)
α Bunsen coefficient (dimensionless)
α_0 Bunsen coefficient of water (dimensionless)
α_{el} Bunsen coefficient of salt solution (dimensionless)
θ exposure time (s)
δ parameter
μ viscosity ($g\ cm^{-1}\ s^{-1}$)
χ association parameter (dimensionless)
ϱ density ($g\ cm^{-3}$)
σ surface tension ($dyne\ cm^{-1}$)
φ age distribution of fluid elements (dimensionless)

10 References

1. Quicker, G., Schumpe, A., Konig, B., Deckwer, W.-D.: Biotech. Bioeng. *23*, 635 (1981)
2. Schumpe, A., Quicker, G., Deckwer, W.-D.: Adv. Biochem. Eng. *24*, 1 (1982)
3. Schumpe, A., Adler, I., Deckwer, W.-D.: Biotech. Bioeng. *20*, 145 (1978)
4. Sechenov, M.: Ann. Chim. Phys. *25*, 226 (1892)
5. Danckwerts, P. V.: Gas-Liquid Reactions, McGraw-Hill, N.Y. 1970
6. Findlay, A., Shen, B.: J. Am. Chem. Soc. *101*, 1459 (1912)
7. Onda, K., Sada, E., Kobayashi, T., Kito, S., Ito, K.: J. Chem. Eng. (Japan) *3*, 18 (1970)
8. Schumpe, A., Deckwer, W.-D.: Biotech. Bioeng. *21*, 1075 (1979)
9. Bird, R. B., Stewart, W. E., Lightfoot, E. N.: Transport Phenomenon, John Wiley, New York 1960
10. Bennett, C. O., Myers, J. E.: Momentum, Heat, and Mass Transfer (3rd Ed.). McGraw-Hill, New York 1982
11. Himmelblau, D. M.: AIChE (Modular Instruction), Stagewise and Mass Transfer *4*, 16 (1983)
12. Davies, G. A., Ponter, A. B., Crain, K.: Can. J. Chem. Eng. *45*, 372 (1967)
13. Ratcliff, G. A., Holdcroft, J. G.: Trans. Inst. Chem. Engrs. *41*, 315 (1963)
14. Nijsing, R. A. T. O., Hendriksz, R. H., Kramers, H.: Chem. Eng. Sci. *10*, 38 (1959)
15. Harriott, P.: Can. J. Chem. Eng. *40*, 60 (1962)
16. Smith, M. D.: M.S. Thesis, Department of Chemical Engineering, State University of New York at Buffalo 1984
17. Tsao, G. T., Lee, Y. H.: Ann. Reports on Fermentation Processes *3*, 74 (1979)
18. Bailey, J. E., Ollis, D. F.: Biochemical Engineering Fundamentals, McGraw-Hill, New York 1977
19. Rushton, J. H., Costich, E. W., Everett, H. J.: Chem. Eng. Progress *46*, 467 (1950)
20. Calderbank, P. H.: Trans. Inst. Chem. Engrs. *36*, 443 (1958)
21. Knoche, W.: Biophysics and Physiology of Carbon Dioxide, (Bauer, C., Gros, G., Bartels, H. Eds.), p. 3, Springer-Verlag, Berlin, FRG 1980
22. Ishizaki, A., Shibai, H., Hirose, Y., Shiro, T.: Agr. Biol. Chem. *35*, 1733 (1971)
23. Roughton, F. J. W.: J. Am. Chem. Soc. *63*, 2930 (1941)
24. Shedlovsky, T., MacInnes, D. A.: ibid. *57*, 1705 (1935)
25. MacInnes, D. A., Belcher, D.: ibid. *55*, 2630 (1933)
26. Wissburn, K. F., French, D. M., Patterson, A.: J. Phys. Chem. *58*, 693 (1954)
27. Otto, N. C., Quinn, J. A.: Chem. Eng. Sci. *26*, 949 (1971)
28. Kernohan, J. C.: Biochim. Biophys. Acta *81*, 346 (1964)

29. Kernohan, J. C.: ibid. *96*, 304 (1965)
30. Tsao, G. T.: Chem. Eng. Sci. *27*, 1593 (1972)
31. Alper, E., Deckwer, W.-D.: ibid. *35*, 549 (1980)
32. Alper, E., Lohse, M., Deckwer, W.-D.: ibid. *35*, 2147 (1980)
33. Donaldson, T. L., Quinn, J. A.: ibid. *30*, 103 (1975)
34. Meldon, J. H., Smith, K. A., Colton, C. K.: ibid. *32*, 939 (1977)
35. Lander, R. J., Smith, D. R., Quinn, J. A.: ibid. *34*, 747 (1979)
36. Matson, S. L., Herrick, C. S., Ward III, W. J.: Ind. Eng. Chem. Process Des. Dev. *16*, 370 (1977)
37. Jones, R. P., Greenfield, P. F.: Enzyme Microb Technol. *4*, 210 (1982)
38. Yagi, H., Yoshida, F.: Biotech. Bioeng. *19*, 801 (1977)
39. Ishizaki, A., Hirose, Y., Shiro, T.: Agr. Biol. Chem. *35*, 1852 (1971)
40. Ishizaki, A., Hirose, Y., Shiro, T.: ibid. *35*, 1860 (1971)
41. Nyiri, L., Lengyel, Z. L.: Biotech. Bioeng. *10*, 133 (1968)
42. Aiba, S., Humphrey, A. E., Millis, N. F.: Biochemical Engineering Academic Press, New York 1965
43. Esener, A. A., Kossen, N. W. F., Roels, J. A.: Biotech. Bioeng. *22*, 1979 (1980)
44. Barford, J. P., Hall, R. J.: ibid. *21*, 609 (1979)
45. Smith, M. D., Ho, C. S.: Chemical Engineering Communications *37*, 2 (1985)
46. Mou, D. G., Cooney, C. L.: Biotech. Bioeng. *25*, 225 (1984)
47. Doyle, M. P.: European J. Appl. Microbiol. Biotechnol. *17*, 53 (1983)
48. Chen, S. L., Gutmanis, F.: Biotech. Bioeng. *18*, 1455 (1976)
49. Hirose, Y., Sonoda, H., Kinoshita, K., Okada, H.: Agr. Biol. Chem. *32*, 851 (1968)
50. Cooney, C. L., Wang, D. I. C., Mateles, R. I.: Biotech. Bioeng. *11*, 169 (1968)
51. Bylinkina, E. S., Nikitima, T. S., Biryukov, V. V., Cherkasov, O. N.: Biotechnol. Bioeng. Symp. *4*, 197 (1974)
52. Pirt, S. J., Mancini, B. J.: J. Appl. Chem. Biotechnol. *25*, 781 (1975)
53. Ho, C. S., Smith, M. D.: Biotech. Bioeng. *28*, 668 (1986)
54. Righelato, R. C., Trinci, A. P. J., Pirt, S. J.: J. Gen. Microbiol. *50*, 399 (1968)
55. Smith, M. D., Ho, C. S.: J. Biotechnology *2*, 347 (1985)
56. Miles, E. A., Trinci, A. P. J.: Trans. Br. Mycol. Soc. *81*, 193 (1983)
57. Pirt, S. J., Callow, D. J.: Nature *184*, 307 (1959)
58. Morton, A. G.: Proc. R. Soc. Lond. B *153*, 548 (1961)
59. Trinci, A. P. J., Righelato, R. G.: J. Gen. Microbiol. *60*, 239 (1970)
60. Collinge, A. J., Miles, E. A., Trinci, A. P. J.: Trans. Br. Mycol. Soc. *70*, 401 (1978)
61. Bent, K. J., Morton, A. G.: ibid. *46*, 401 (1963)
62. Albert, B., Bray, D., Lewis, J., Raff, M., Roberts, K., Watson, J. D.: Molecular Biology of the Cell, Garland Publishing Company, New York 1983
63. Houslay, M. D., Stanley, K. K.: Dynamics of Biological Membranes, John Wiley and Sons, New York 1982
64. Stewart, P. R., Rogers, P. J.: "Fungal Dimorphism", in Fungal Differentiation, (Smith, J. E. Ed.), Marcel Dekker, Inc., New York 1983
65. Jaffe, L. F.: Phil. Trans. R. Soc. Lond. B. *295*, 553 (1981)
66. Harold, F. M., Kropf, D. L., Caldwell, J. C., Exp. Mycol. *9*, 183 (1985)
67. Jaffe, L. F., Muccitelli, R.: J. Cell Biol. *63*, 614 (1974)
68. Kropf, D. L., Lupa, M. D. A., Caldwell, J. H., Harold, F. M.: Science *220*, 1385 (1983)
69. Kropf, D. L., Caldwell, J. H., Gow, N. A. R., Harold, F. M.: J. Cell Biol. *99*, 486 (1984)
70. Ho, C. S., Shanahan, J. F.: CRC Critical Reviews in Biotechnol. *4*, 185 (1986)

Conservation of Yeasts by Dehydration

Mártin J. Beker and Alexander I. Rapoport
August Kirchenstein Institute of Microbiology,
Latvian Academy of Sciences Kleisti 226067, Riga/USSR

1 Introduction 128
2 Anabiosis as a Biological Phenomenon 129
2.1 Anabiosis in Free Nature 129
2.2 Resistance of Dried Microorganisms to Extreme Environmental Factors 129
2.3 Duration of Yeast Organism Anabiosis 130
3 Dehydration of Yeasts 131
3.1 Water and Its Transport in Yeast Biomass During Drying 131
3.2 Resistance of Yeasts to Drying 132
3.3 Effect of Cultivation Conditions on Viability of Yeasts after Drying 133
4 Structural Changes of Yeast Cells During Dehydration, Rehydration and Reactivation 134
4.1 Structural Changes of Cells During Dehydration 134
4.2 Rehydration and Reactivation of Cell Structures 141
4.3 Separation and Elimination of Damaged Parts of Cells. Membrane Formations 143
4.4 Permeability of Cells after Drying Rehydration 144
4.5 Changes of Intracellular Yeast Membranes by Drying-Rehydration 146
5 Biochemical Changes in Yeast Cells During Drying, Rehydration and Reactivation 147
5.1 Changes of Nucleic Acids 147
5.2 Effect of Drying on Protein Compounds 149
5.3 Changes in the Composition of Lipids in Cells 152
5.4 Changes of the Carbohydrate Content in Cells 154
5.5 Polyphosphate Changes in Cells 155
6 Restoration of Yeast Populations 156
6.1 Rehydration 156
6.2 Reactivation of Yeasts 158
6.3 Restoration of the Growing Population 160
7 Application of Dehydration for Production of Active Dry Yeasts 162
8 Conclusion 164
9 References 165

The presented material concerns the theoretical basis for obtaining high-quality active dry biopreparations. It deals with the present understanding of anabiosis, contains data on yeast resistance against dehydration and the limits for preserving the viability of microorganisms in anabiosis. The process of water transport in yeast biomass during dehydration is discussed.

The changes and transformations in yeast cells occuring after their transition into anabiosis are described. The main intracellular reactions in view of the maintenance of yeast viability after dehydration and subsequent reactivation are shown. A hypothetic model of the possible transformations in separate parts of biomembranes is developed and data on the effect of dehydration on the main intracellular compounds — nucleic acids, proteins, lipids, carbohydrates and polyphosphates are presented. Finally the effects of rehydration conditions on restoration are demonstrated and the main methods for obtaining dry active baker's yeasts for commercial purposes are described.

Advances in Biochemical Engineering/
Biotechnology, Vol. 35
Managing Editor: A. Fiechter

1 Introduction

One of the most remarkable natural phenomena — anabiosis — was discovered in 1701 by the outstanding naturalist-microscopist of the 18th century — Anthony van Leeuwenhoek. Though this phenomenon has been studied in various parts of the world for about three hundred years, substantial progress in this field was made in the second half of the 20th century only.

At present the main task is to understand more profoundly the basic processes taking place in the cells of living organisms after transition into anabiosis and subsequent recovering of their normal viability.

The present review deals with these phenomena using yeast as a model of yeast organisms.

The term 'anabiosis' was coined by Preyer [1] at the end of the 19th century in order to describe outwardly "lifeless yet at the same time viable living organisms". This term has rooted itself in the literatures of many countries, the USSR included [2]. Keilin suggested the term 'cryptobiosis', that is "the state of an organism when it shows no visible signs of life and when its metabolic activity becomes hardly measurable or comes reversibly to a standstill" [3]. All over this review we shall use the term 'anabiosis'.

Theoretically, in laboratory scale experiments, an extensive drying of living organisms may lead to a complete suspension of metabolic processes (at least of those that can be registered by the available methods and techniques) [4]. On the other hand, some works [5,6] revealed that dried organisms still use certain, though very small, amounts of oxygen. Oxygen consumption rapidly increases with the regained moisture content in cells.

Such discrepancies in the results may first of all be due to the different levels of drying of the objects under study.

When the microorganisms in natural environment dry up, the moisture content of cell biomass, in keeping with the rules of equilibrium, is not lower than 3—6%, most often it is 6–15%. It was demonstrated that with an 8–12% moisture content in yeast cells, corresponding to the state of anabiosis, a gas exchange takes place between dry baker's yeast and air in tightly corked flasks. After 20 days of storage the content of carbon dioxide in the atmosphere reached 3.9%, but oxygen content decreased to 16.4% [7].

Our opinion is that at present anabiosis should be understood as a reversible state of biological systems, when metabolism is extremely low or suspended.

Depending on the ways of transition into anabiotic state, Keilin [3], followed by other investigators [8], distinguished between:

1 — anhydrobiosis — a state, caused by significant losses of water through evaporation;

2 — cryobiosis, taking place by freezing living organisms;

3 — osmobiosis — a state, brought about by extraction of water from the organisms by various solutions with a high osmotic pressure;

4 — anoxybiosis, caused by a decrease of oxygen concentration in the gas phase below critical limits for aerobic metabolism.

2 Anabiosis as a Biological Phenomenon

2.1 Anabiosis in Free Nature

Anabiosis is very widely spread in nature. Some species of rotifiers, tardigrades, nematodes, larvae of some insects, eggs of some crustaceans, several other representatives of the fauna on our planet, as well as plant seeds and microorganisms are capable to reactivate a "dormant" metabolism. Various aspects of anabiosis in these organisms are discussed in detail in a number of reviews and books [4,8–13].

In microorganisms the state of anabiosis can be expressed in two ways. Under unfavourable conditions, many species are known to display a series of special reactions, to change their metabolism to a considerable degree and to produce spores. Most often the sporulation is due to a nutritional deficiency. This phenomenon has been the object of active studies in many countries, and a great number of reviews and experimental reports have been dedicated to this problem [14–17].

Drying does not usually stimulate spore formation. Under these conditions it is the vegetative cells of microorganisms that pass into the state of anabiosis. In nature the microorganisms in air, on plant surface or in the soil can very often be subjected to drying up, stay in anabiosis, rehydrate and revitalize to normal activity.

Our review is an attempt to discuss and sum up the available data on various aspects of yeast anabiosis after dehydration, and its practical importance for industry.

2.2 Resistance of Dried Microorganisms to Extreme Environmental Factors

Studies on the various aspects of anabiotic state of living organisms have detected their unusual resistance to the extreme environmental factors. Dried organisms were found to be more stable (as compared with organisms in the state of active viability) to high and low temperatures, large doses of UV- and X-rays or superhigh vacuum. The results of these studies are presented in a number of reviews [2–4,8,9,18].

With regard to living systems in the state of anabiosis Lozina-Lozinsky distinguishes between their physiological and physical resistance [19]. He understands the physiological resistance as the resistance of a functional biological system connected with adaptive and reparative processes. Thereby the physical (structural) resistance of biological objects is expressed by a total absence of vitality.

From the same point of view, we may conclude that in case of a complete suspension of metabolism (when using special drying methods) integrity of the organism depends on the physical resistance (durability) of biological structures. As natural and close to natural laboratory conditions for drying living organisms do not ensure a complete inactivation. The viability of a given object is affected by various detrimental factors depending on structural and physiological resistance. In this case it is important that the repair systems of the organism are intact, recovering from demaged cell ultrastructures or macromolecules.

2.3 Duration of Yeast Organism Anabiosis

Determining the critical duration of the anabiotic state of living organisms including yeasts, one has to cope with serious methodological problems. It is not easy to find a suitable model for such investigations. Also only few data are available on viability of anabiotic yeasts.

Viable yeasts *Saccharomyces cerevisiae* and *Rhodotorula pallida* were found in food concentrates left behind by the second expedition of Captain R. Scott in 1911 and having remained under the snow for 50 years [20].

Studies on dry soil from plant roots in botanical collections showed viable fungi in samples after 100 years of storage [21].

Some papers report also on yeast storage at low temperatures — from several hours up to 5 years [22–25].

Very interesting results have been obtained by S. S. Abizov et al., who studied the samples of ice probes, obtained during the 20th, 21st, 22nd and 25th Soviet Antarctic expeditions. Viable microorganisms were detected in the depth of the glacier in the Central Antarctic region, in the vicinity of the "Vostok" station where the ice temperature is 55–57 °C below zero. Both, sporulating and non-sporulating bacteria (e.g. actinomycetes) were detected as well as yeasts and mycelial fungi [26,27].

The data obtained show that viable yeast cells can be found in 3,250 years old glacier horizons and the other viable microorganisms were even obtained from 12,000 years old horizons [26].

The theoretical estimate of the possible limit for maintaining microorganism viability in anabiotic state was made [28], using the theory of the absolute rates of chemical reactions [29] and experimental data on the duration of sterilization process depending on the temperature used. It was established that E_{act} at which the damages in microorganism spores, detrimental for their viability, may occur by heat dosages of 32.8–40.7 kcal mol^{-1}. On the basis of the obtained values Aksyonov calculated the maximum preservation periods of microbial viability in the state of anabiosis. As to these data when stored at 20 °C such organisms would perish after 10^4–10^6 years [28].

It is interesting to note, that this period of time, according to Aksyonov correlates quite well with the results of the studies on amino acids isolated from geological deposits [30,31]. These investigations demonstrated that the process of L-amino acid racemization takes place during approximately up to 100,000 years eventually yielding in an optically inactive mixture of L- and D-amino acids. It is also evidenced that some amino acids (arginine, hystidine, serine, threonine and cysteine) may decompose at ambient temperatures during 10^5–10^7 years [30,32].

Thus, the above results indicate that the persistence of the anabiotic state at temperatures around 20 °C is still rather limited. The data [28] show simultaneously that at low temperatures (—20 to —30 °C) this period may increase significantly and reach tens to thousands of millions of years. Aksyonov however, points out that some "limiting factors" like ionizing radiation may reduce this period drastically.

3 Dehydration of Yeasts

3.1 Water and its Transport in Yeast Biomass During Drying

Microbial biomass contains 70–90% of water. Biopolymers and membranes are dispersed in a water medium, and life takes place only in the presence of water. Water is a structural component of both, biopolymers and biomembranes. Besides that, water as a substance is directly involved in a number of biochemical reactions.

The question of water structure is still disputable and it has been discussed in great detail in a number of reviews [33–36]. The term 'bound water' has been used in biological literature since long ago. Today bound water is understood as that part of intracellular water, which combines directly with proteins, nucleic acids, membranes or other substances and which is responsible for maintaining their structural organization [37–41]. Yeast biomass contains 15–20% of bound water, cell biomembranes about 25% [42].

Convective drying of yeast biomass refers to two processes:

1) water from the surface of the outer layers of biomass granules evaporates into the atmosphere (intergranular water)
2) within the granules water displaces by way of diffusion (intragranular translocation)

Evaporation rate can be expressed as follows:

$$I = -D \frac{P_b}{P_b - P_s} \Delta C ,$$

where

I — the amount of evaporated liquid in a time unit from a surface unit, kg m^{-2} h^{-1};
D — coefficient of diffusion, m^2 h^{-1};
ΔC — concentration gradient, kg m^{-4};
P_b — atmospheric pressure, mm Hg
P_s — partial water vapor pressure, mm Hg

Intracellular structures form an internal system of capillaries and pores. Therefore during cell drying the moisture travels by way of diffusion in the form of liquid and vapour.

The law of moisture conductivity (i) for dehydration of microbial biomass can be expressed as follows [43]:

$$i = -k\gamma_0\Delta_u ,$$

where

i — water conductivity, kg m^{-1} h^{-1};
k — coefficient of moisture conductivity of biomass, m^2 h^{-1}, depending on coefficients of vapour and liquid diffusion;
γ_0 — mass of the absolute dry body per unit of the moist body volume, kg m^{-3};
Δ_u — moisture gradient.

During convective drying a rapid linear decrease of the moisture content occurs first (Fig. 1). Then it asymptomatically approaches the moisture equilibrium.

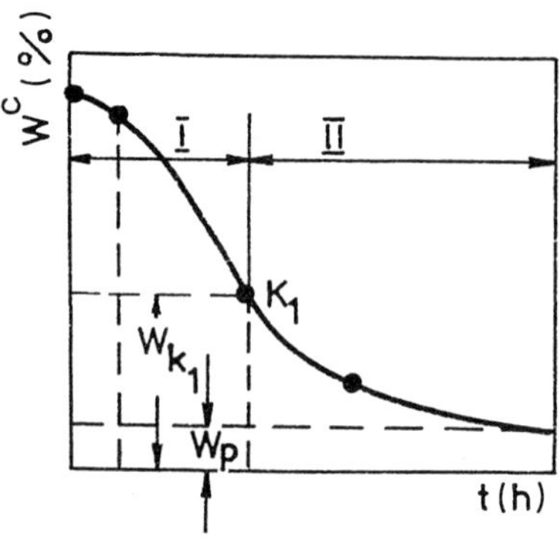

Fig. 1. The curve of microorganism biomass drying. W_{k_1} — bound water content, %, W_p — moisture content in biomass after drying, %. Period I characterizes the isolation of free water from cells. Period II characterizes the isolation of bound water from cells

During the period of a constant drying rate the intercellular and a part of intracellular (free) water is evaporated. When the bound water starts to evaporate, the drying rate decreases. This takes place in baker's yeast at about 20% moisture [44]. The data on bound water contents in yeast cells are confirmed by differential scanning microcalorimetry [45].

In yeast cells the state of water at various moisture contents was studied by measuring a number of physical parameters, such as pressure of water vapour, heat evaporation and nuclear magnetic resonance spectrum [46]. It was established that up to 20% moisture content water fills up the void volume and acts as a solvent. It is the author's opinion that biochemical reactions are still possible at moistures exceeding 20%, even at low rates. At a 20% moisture content however the physiological behaviour of cells drastically changes. In the span between 20 and 10% moisture water together with dissolved substances forms a gel and can no longer act as a medium for biochemical reactions. A further decrease of moisture leads to the translocation of water in cells in the form of vapour and changes the physical parameters of yeasts rather essentially.

Aksyonov [47,48] has obtained interesting data when studying dried yeasts by NMR. He found "isolated preserved water" in an amount which depends on the phase of growth and makes up 2.5–4.0% of the biomass. Such isolation of water in dried viable microorganisms is caused by changes of membrane protein conformation taking place after drying, which, in their turn, change the permeability of intracellular membranes. It is supposed that water is isolated in pockets formed by the lipid surroundings; the latter, together with other membrane components isolate this water from external effects of solvents other than water, ensuring a rather high stability of the system with increased resistance to drying the preserved water in dried cells of yeast tends to accumulate.

3.2 Resistance of Yeast to Drying

Resistance against drying is a specific characteristic of microorganisms and depends to taxonomical and physiological properties determining the chemical composition of the species or strains. For example, *Saccharomyces cerevisiae 14* is more resistant than *Saccharomyces cerevisiae Tomskaya 7* [49]. The same strain is also less sensitive to freeze-drying as compared with *Saccharomyces cerevisiae VKM Y-375*, *Saccharomyces cerevisiae VKM Y-381* and *Saccharomyces cerevisiae VKM Y-398* [50].

Interesting data were observed in the studies of bacterial anabiosis. At moisture content of cells within 20–30% some bacteria show an increased lability. For example up to 99.5% of *Lactobacterium acidophylum* cells are inactivated in 24 h at this moisture level in contrast to yeasts where no such phenomena were detected [18].

In order to establish the optimum water content of dried microorganisms for storage, viability during long periods of time must be considered. It is possible, for example, to bring *Saccharomyces cerevisiae* down to 15–20% moisture practically without losses of viability. However, a considerable inactivation can be observed under these conditions during storage. Yeasts with a 12–13% water content are not suitable for storage either, whilst an 8–10% moisture allows viability a remarkable period of time [18].

3.3 Effect of Cultivation Conditions on Viability of Yeasts after Drying

Cultivation conditions greatly influence microorganism resistance to drying. The resistance of one and the same strain may vary within very wide limits, depending on the composition of nutrient medium used for cultivation. For example, viability of *Saccharomyces cerevisiae*, grown on a rich molasses medium may reach 90%, while that of the same strain grown on a synthetic medium is only 20–40% [18].

It is known that a high viability of dried yeast can be reached if grown on concentrated molasses media with a limited nitrogen content [49,51,52].

It was noted also that yeasts sampled at the exponential growth phase are considerably more labile against drying than those of the stationary growth phase [49]. This phenomenon is typical for cell material from complex as well as from synthetic media.

The great importance of nutrient supplementation for preserving yeast viability was established with batch and continuous cultivations [52]. Also, rich molasses media do not necessarily provide resistant cells when dried. This may occur due to an overdose of nitrogen. Yeasts with 6.4% of nitrogen (dry weight) lost about 15% of viability after drying, but those with 9.8% were practically totally inactivated [51]. The increased nitrogen content from 6.3 to 8.7% in the medium induced the protein content of the biomass of 39.7 instead of 32.6%. Also, a total RNA raised from 5.2 to 10.2% [53]. It is assumed simultaneously that the amount of macromolecule-bound water is also increasing. During drying of yeast to the "standard" moisture content the amount of bound water to be extracted increases. This of course causes irreversible damages of macromolecules [18].

By enriching the synthetic medium, it is possible to increase yeast stability for drying. Addition of 0.1% of yeast extract increased viability from 30 to 66% [54]. Vitamins (the most effective was biotin) increased viability approximately by 30% [54]. Also addition of oleic increased cell resistance by 10% [54]. Glutamic, asparaginic acids and glutatione addition increased the resistance of convectively dried yeasts by 54, 23 and 24% respectively [54].

It is known that the resistance of yeasts against drying can be considerably raised by increasing the trehalose content in cells. Its synthesis is enhanced at elevated temperature [55] and decreasing aeration at the final growth stage [56]. Furthermore trehalose content in *Saccharomyces cerevisiae* up to 15–16% can be obtained at increased osmotic pressures [56] or cell starvation of 2–3 h prior to termination of the batch [56].

A special treatment after cell growth under high osmotic pressure can greatly increase the resistance of yeast cells in the process of subsequent rehydration (by 5–64%) [57].

Thus, the presented data demonstrate the importance of cultivation conditions or "pretreatment" prior to dehydration.

4 Structural Changes of Yeast Cells During Dehydration, Rehydration and Reactivation

4.1 Structural Changes of Cells During Dehydration

Light-optical studies of yeast cells during dehydration revealed that mild conditions (on a thin wort-agar film or in a flow of warm air) do not affect the cell morphology [50]. Drying on glass at 25 °C a slight decrease in size took place, the cytoplasm became granular, the number of lipid bodies increased; finally the cells changed their form and became elongated [58]. Changes of cell form were observed also after freeze-drying of several yeast strains, less resistant to this procedure [50]. Plasmolysis

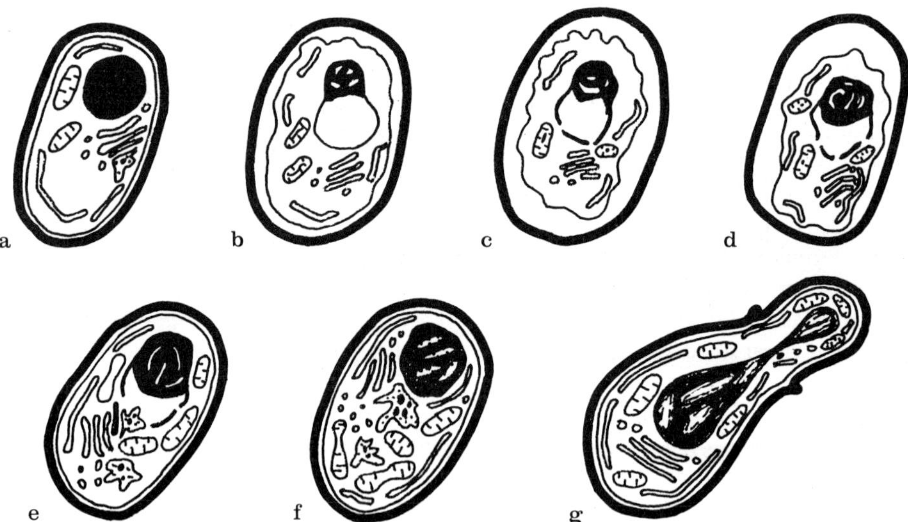

Fig. 2a–g. A scheme of several structural transformations in a yeast cell during dehydration (**a–d**) and subsequent reactivation (**e–g**) **a**: initial cell, **b–d**: formation of cytoplasmic membrane invagination during dehydration, plasmolysis, chromatin condensation and separation of chromosome containing part of the nucleus, **e–f**: moistening of cells during reactivation, a further degradation of the part of the nucleus to be separated and its elimination by phagosomes, **g**: division of a reactivated cell

took place and autolysis of the structural components occured. In the majority of cases only oval cells (with an unaltered morphology) preserved their viability.

The use of NaCl and glycerin solutions at the initial stages of cell dehydration revealed changes of the cytoplasm and the macromolecule-binding acridine dyes [59,60].

An electron microscope study showed chromatin condensation in cell nuclei during the early stages of dehydration. In *Candida utilis* condensed chromatin was usually localized at one pole of the nucleus. This process was mostly accompanied by separation of the homogeneous chromosomeless part of the nucleus, resulting in an ingrown nuclear membrane (Fig. 2). The pores of the nuclear membrane became very wide and the nucleoplasm was resorbed gradually [61]. It is possible that this phenomenon is connected with the activation of protection mechanisms for the preservation of DNA, manifested by a dense packing of chromatin. Also a decrease of nucleus volume on account of the nucleoplasm containing a minimum of water was observed [61–63].

Comparison of the nuclear ultrastructure of different strains revealed various levels of chromatin condensation. Spiralized chromosomes are typical for dehydrated cells in populations of high viability [64,65].

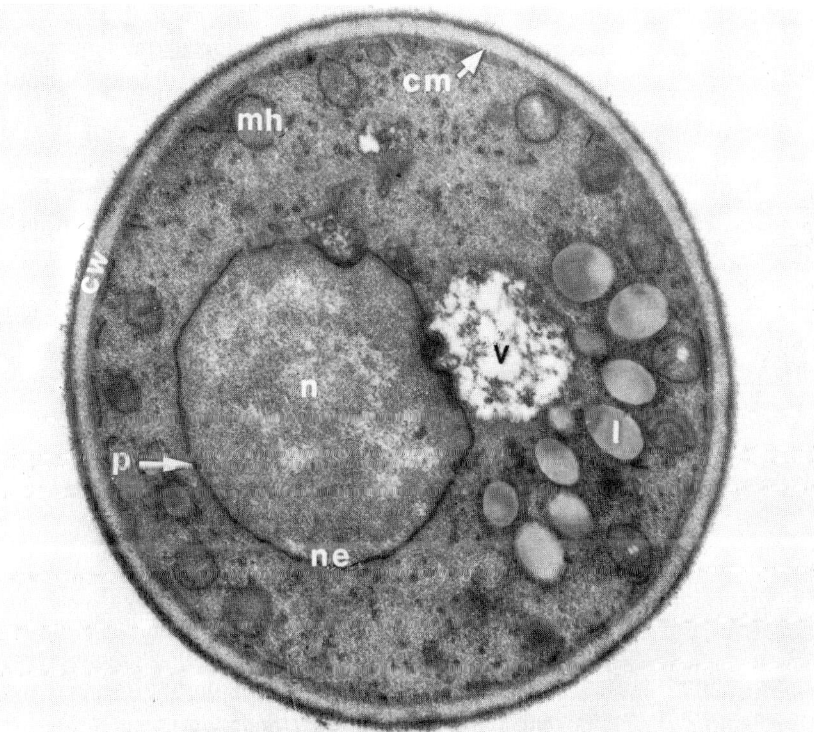

Fig. 3. Electron microscopy of ultra-thin sections of yeast *Saccharomyces cerevisiae* cells. Intact cell. cw — cell wall, cm — cytoplasmic membrane, n — nucleus, ne — nuclear envelope, p — nuclear pore, mh — mitochondrium, l — lipid body, v — vacuole, M 19.000

Chromatin condensation was obvious when searching for a method of light-optical visualization of chromosomes. In this case yeast protoplasts were subjected to drying. The authors succeeded in obtaining various levels of chromatin condensation in the cells [66].

It was also established that not only chromatin condensation in the nuclei takes place, but also nuclei may change their form and the sorptive properties of nucleoplasm (with regard to acridine fluorochromes) may become destroyed [67]. In a number of cases fragmentation of cell nuclei was caused [50]. Electron microscopy of ultra-thin sections of dehydrated yeast cells shows a notable widening of pores in nuclear membrane (Figs. 3, 4) [68,69]. Mitochondria or lipid bodies were often localized around such "open" parts of the nucleus [68-70]. In analogy data were obtained, when some other physical and chemical factors affected the cells [71,72]. Transformations detected in dehydrated cells indicate a weakening of nuclear membrane rigidity. This is confirmed by freeze-etching data of dehydrated cells. Considerable contraction of nuclear pores can be detected [73]. A notable increase of pore size in dried cells by partial hydration during the fixation of specimen and subsequent treatment for ultra-thin sectioning indicates a weakened rigidity of nuclear membranes.

Scanning electron microscopy established that dried yeast organisms use to flatten (Figs. 5, 6). A fraction of cells have either a folded or rough surface after dehydration (Fig. 7). Sometimes protrusions or bubble-like, drop-like particles of various sizes and forms and whole clusters of them may appear (Fig. 8) [18,74,75]. A parallel computerized optico-structural analysis of the geometrical parameters of dried and subsequently rehydrated yeast cells demonstrated elongation after drying.

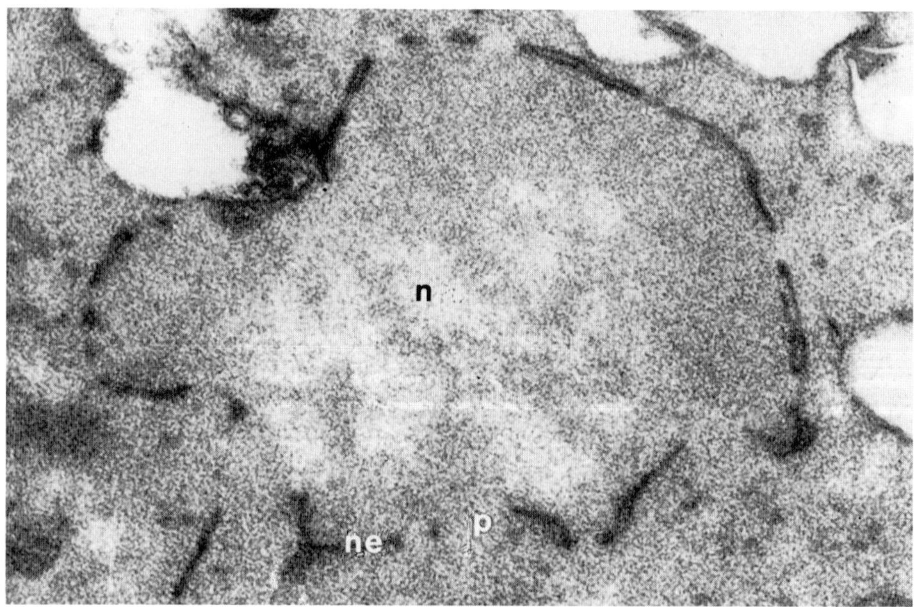

Fig. 4. Electron microscopy of ultra-thin sections of yeast *Saccharomyces cerevisiae* cells. Nucleus (n) of a dehydrated cell with widened pores (p) of nuclear envelope (ne). M 55.000

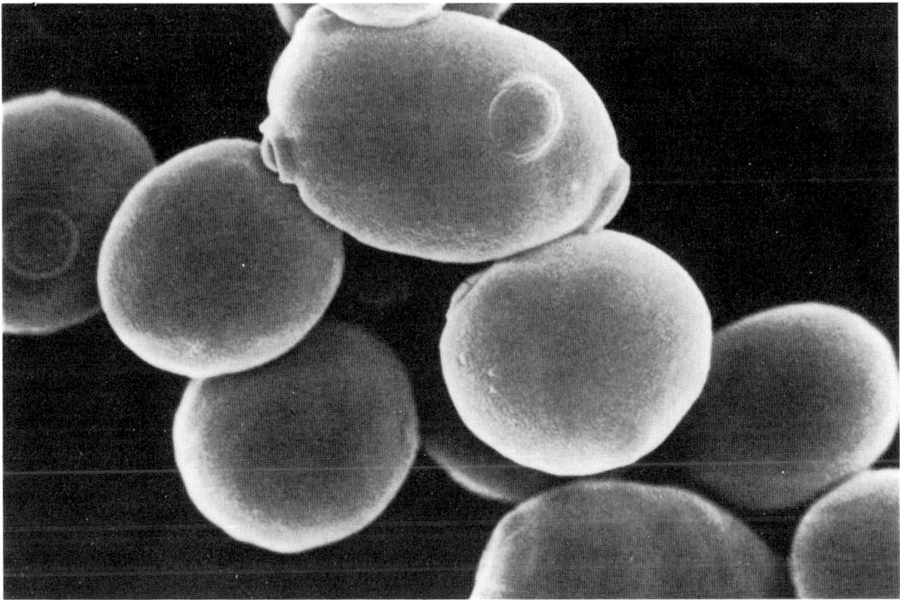

Fig. 5. Scanning electron microscopic of yeast *Saccharomyces cerevisiae* cells. Intact cells. M 7000

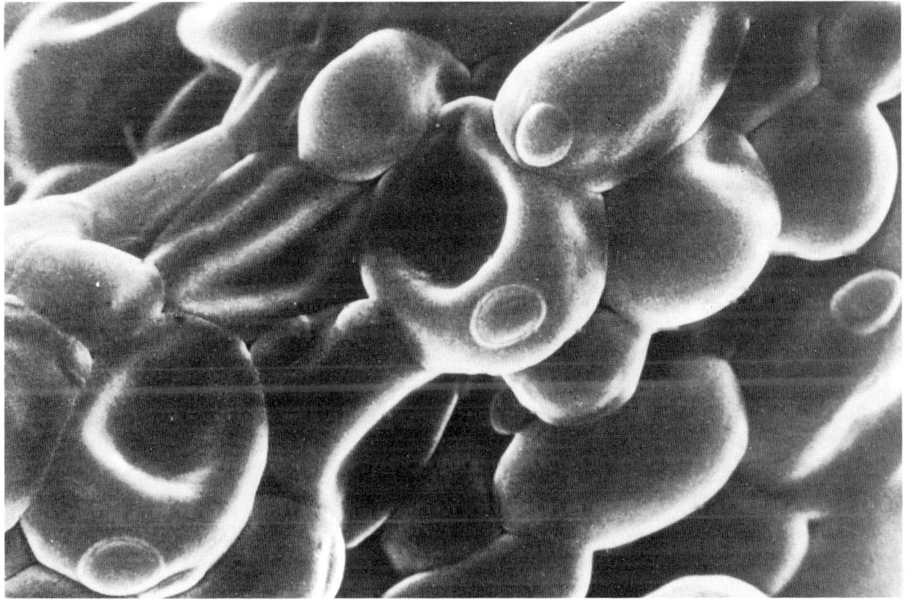

Fig. 6. Scanning electron microscopy of yeast *Saccharomyces cerevisiae* cells. Yeast cells after dehydration. M 5000

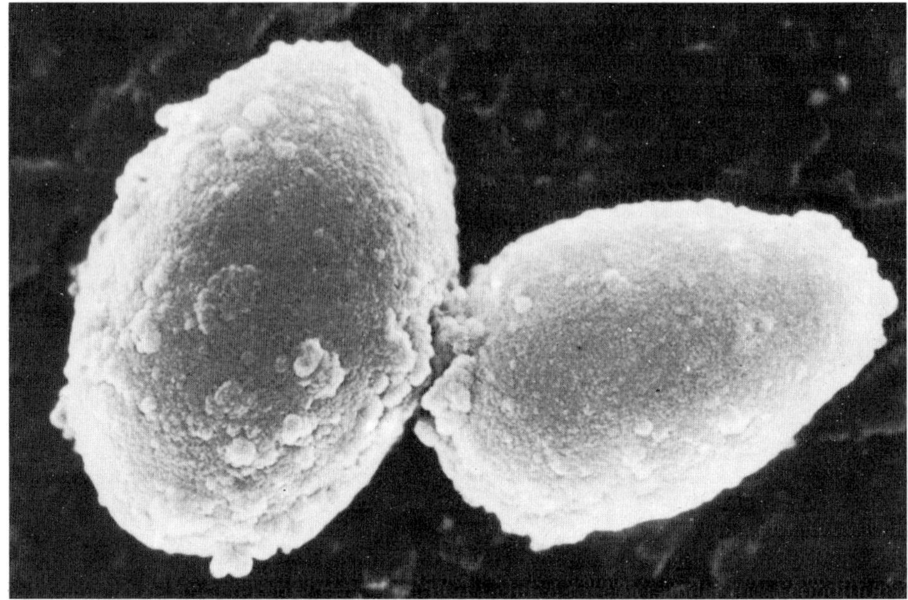

Fig. 7. Scanning electron microscopy of yeast *Saccharomyces cerevisiae* cells. Dehydrated cells after rehydration. M 16.000

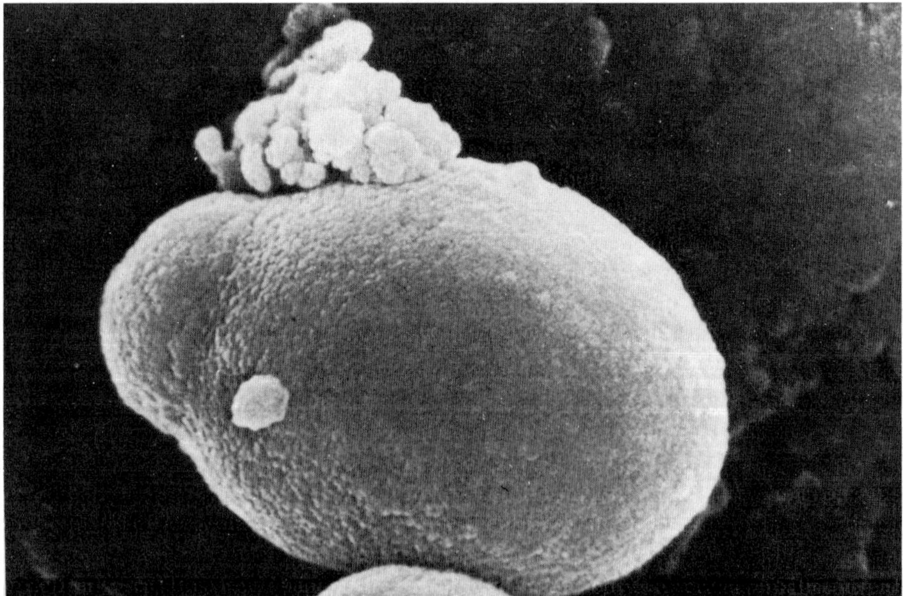

Fig. 8. Scanning electron microscopy of yeast *Saccharomyces cerevisiae* cells. Dehydrated cell after rehydration. M 22.000

Fig. 9. Electron microscopy of *Saccharomyces cerevisiae* cell surface after staining with metal. Smooth surface of intact cell. M 80.000

Fig. 10. Electron microscopy of *Saccharomyces cerevisiae* cell surface after staining with metal. Branched fibrillas on the surface of dehydrated cell. M 80.000

Restoration of their initial form and size appeared only gradually during reactivation [76]. It was concluded that certain glucane component of yeast cell walls changed during drying.

Another method of electron microscopy — metal shadowing — showed the presence of branched mannane-proteid structures on the surface similar to yeast flocculation (Fig. 9, 10). This refers to changes in the structural organization of mannane-proteid elements of cell wall after drying-rehydration. Staining with alcyane blue demonstrated a notable increase of the negative charge of the surface by dehydration [77]. These transformations of cell walls may appear as responsible for the frequent aggregation to large conglomerates of tens of cells which cannot easily be detached from each other [75,78].

The cytoplasmic membrane has deep invaginations, obviously due to a shrinkage of the cells by dehydration (Fig. 11) [68, 73, 79, 80,].

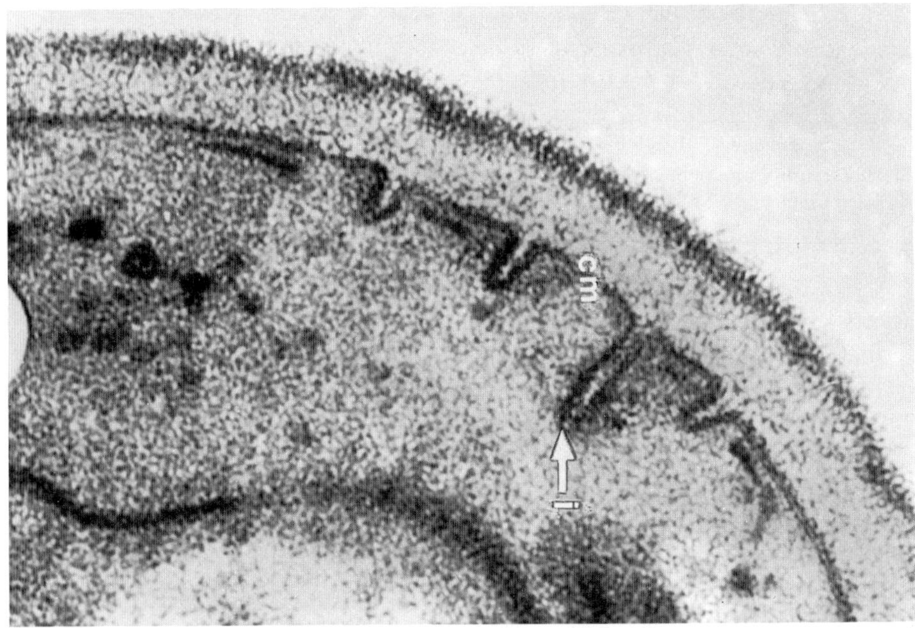

Fig. 11. Electron microscopy of ultra-thin sections of yeast *Saccharomyces cerevisiae* cells. Cytoplasmic membrane (cm) invaginations (i) in a dehydrated cell. M 170.000

The structure and form of mitochondria of dried yeast do not change practically, while mitochondrial DNA becomes more clearly visible [61].

Certain changes were detected also in the vacuoles. When dehydrated, big vacuoles diminish drastically in size and/or become partly fragmented [81]. Since the regulation of osmotic pressure in cells is one of the functions of the vacuole system the diminishing of big vacuoles causes a flattening of cells during drying. An increase of lysosome number in dehydrated cells as shown by luminiscent microscopy can also be connected to the fragmentation of vacuoles [67].

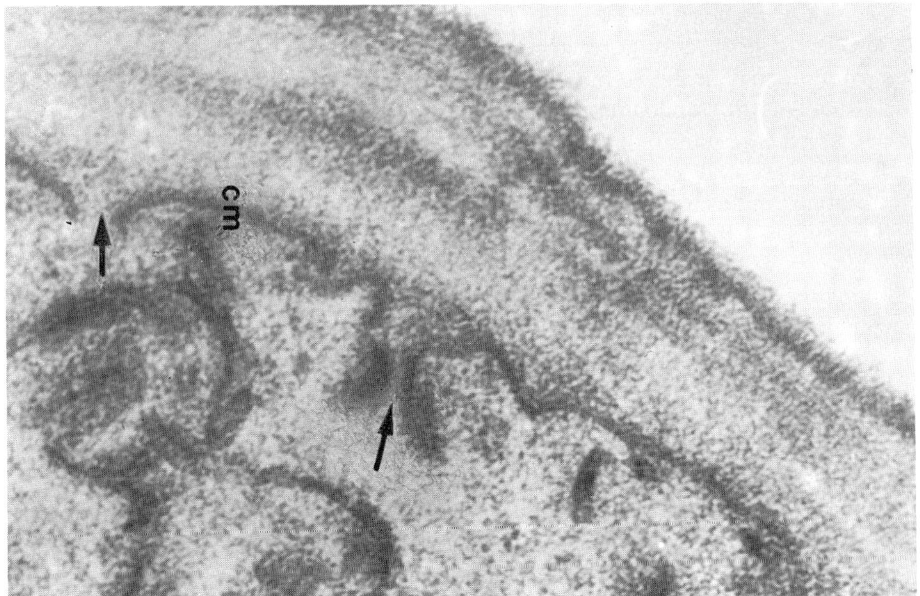

Fig. 12. Electron microscopy of ultra-thin sections of yeast *Saccharomyces cerevisiae* cells. Ruptures (↘) of the cytoplasmic membrane (cm) of a dehydrated cell. M 145.000

Cells inactivated by dehydration differ from those in anabiosis by plasmolysis, showing many ruptures in nuclear and cytoplasmic membranes (Fig. 12). Furthermore, damaged membranes of the endoplasmic reticulum and drastic size decrease of the nucleus are typical [18].

4.2 Rehydration and Reactivation of Cell Structures

Electron microscopy investigations of rehydration process show that slow and gradual rehydration of dried cells from the stationary growth phase in water vapour diminishes membrane damages as compared to a rapid moistening, by just immersing dried yeasts into liquid medium. The viability of rehydrated populations increases, too. This indicates the significance of the rehydration process in preserving the structural integrity of the cell. This factor becomes significant only for membranes reversibly changed during dehydration. This is supported by the fact that in dehydrated cells from the exponential growth phase, with an almost 100% budding, it is impossible to diminish membrane damages during rehydration [82].

Staining of rehydrated viable cells with a strongly diluted solution of neutral red demonstrated a diffuse pink-red cytoplasm color, disappearing in about 2–3 h after reactivation. This indicates reversibility of paranecrosis after drying [18]. Non-specific changes take place in cells at paranecrosis. Further decrease of colloid dispersion and of cytoplasmic pH plus release and excretion of potassium and phosphate ions from cells are typical. Simultaneously, sodium and chlorine ions are introduced into the cells from the medium [83].

Fluorochromation of rehydrated yeasts demonstrated a gradual diminishing of lysosome number during reactivation [67]. The increase of their number during dehydration may be due to the necessary release and "digestion" of various products of disintegration. Analogous processes including increase of lysosome activity can be observed during the first hours of reactivation. In addition the number of damaged cell structures and macromolecules to be released gradually diminishes during restoration, and as a result the number and activity of lysosomes decrease.

It is assumed that increasing lysosome activity is one of the reasons for a partial inactivation of microorganisms during dehydration and subsequent reactivation. This may be connected with hampered permeability or functional integrity of lysosome membranes during dehydration enhancing release of hydrolytic enzymes into cytoplasm and causing autolysis of the cell [18].

It was also shown that during rehydration anabiotic yeasts release potassium and magnesium ions, localized in yeast vacuoles exerting lysosomal functions. It was concluded that the permeability of cytoplasmic and vacuolar membranes for the ions increases. Since the maximum amount of ions is released already during the hydration of cells dehydrated to 20% moisture (Fig. 13), it was supposed that such membrane damages cause the loss of free water [84]. Further studies on the changes of vacuolar membrane permeability would undoubtedly be of great interest.

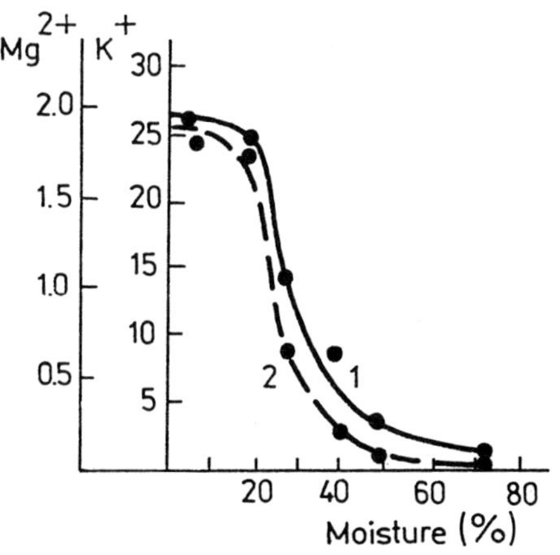

Fig. 13. Release of potassium and magnesium ions during yeast cell rehydration in dependence on the moisture content (μg mg^{-1} dry wt.).
1 — release of potassium ions;
2 — release of magnesium ions

Electron microscopic investigations demonstrated that the functional activity of mitochondria is restored prior to that of other organelles. This is, most probably, due to their well-preserved structural integrity [68]. Analogous results were also obtained from the reactivation of cells treated with ionizing radiation [85]. A rather notable increase of mitochondria indicates an intensified process of reactivation. These conclusions, based on cytological and ultrastructural studies correlate with the necessity of intensified energy metabolism during the reparation of heat-damaged cells of *Staphylococcus aureus* [86], *Salmonella typhimurium* [87] and freeze-dried *Escherichia coli* [88].

Structure and form of the nucleus are normalized later than mitochondria. For some time the preparations may contain cells with "open" nuclear pores. When studying the reactivation of *Candida utilis* culture in which the chromosome spiralization and separation of the homogeneous chromosomeless part of the nucleus was observed, it was demonstrated that the chromosome containing parts of the nucleus increased gradually. In contrast the separated parts were eliminated from the cytoplasm and captured by the phagosomes. In some reactivated cells big vacuoles can be observed, which accumulate and eliminate damaged cell structures from the cytoplasm [61].

The restoration of a normal structure and permeability of cytoplasmic membrane is slower. This process, however, does not proceed simultaneously along the whole cell surface [18].

In the beginning of the reactivation the changes of the form and surface of yeast cells appear more marked with regard to the dehydrated organisms. No correlation could be found between the level of viability of cells, reactivated after dehydration, and the character of the changes in their form and surface. Thus, by far not in all the cells, taken for drying from the exponential growth phase and distinguished by an extremely low viability, it was possible to detect changes of cell form and surface structure, typical of the bulk of organisms.

Bubble-like, drop-like forms and their clusters appeared after rehydration due to the release of various substances into the medium including inorganic compounds, amino acids, nucleotides, carbohydrates, lipids, peptides and even proteins [75]. It was also demonstrated that *Escherichia coli* cells show peculiar bubbles and vesicules when heated. Chromatographic assay of the material excreted let the authors conclude that the composition has much in common with the chemical nature of the outer membrane [89]. The experiments with dehydrated yeasts allow the conclusion that the excreted products are resembling the material of the cell wall and periplasmic space.

Interesting are data on the detected various surface changes including folding of cell wall even after two-hour reactivation (their lag-phase lasts about 3 h). This indicates quite serious changes of the structural organization of the cell wall including glucane component and their prolonged reparation/reactivation time [75].

The changes of cell wall, cytoplasmic membrane, cytoplasm, mitochondria, lysosomes, nucleus were found to be reversible and reactivation of the cell can be observed.

4.3 Separation and Elimination of Damaged Parts of Cells. Membrane Formations

Phase-contrast microscopy of the yeast *Endomyces magnusii* revealed a marked separation of the cytoplasm into viable and irreversibly damaged parts. The viable part resembled that of a normally developing cell. Furthermore it started to elongate, divide, separate by a septation from the newly-formed daughter-cell. Along with this, the structures of the irreversibly damaged part were gradually subjected to autolysis [90]. The frequency of such cell damages reached 2.5–3.0%.

It is assumed that a partial elimination of the damaged parts of the cytoplasm is characteristic only of cells with intact nuclei in mono- or multinuclear yeasts.

There are available data on the segregation of small parts of the cytoplasm found after radiation, by the membranes of the Golgi apparatus [91]. In our case it was cytoplasmic membrane that was responsible for the segregation and the elimination of a considerable part of the cell. The separation of the viable and dying parts of the cell appeared during dehydration [90].

Cytological studies on yeast reactivation from anabiosis showed invaginations of the cytoplasmic membrane into the periplasm. At times mesosome-like formations were found in the periplasmic space and the cytoplasm. It is assumed that these phenomena may be related to the segregative function of the cytoplasmic membrane and other membranes, separating the irreversibly damaged ultrastructures and macromolecules from otherwise functional structures. It is also possible that the accumulation of membrane material is due to a deregulation of cell growth and membrane biosynthesis in cells during restoration [80, 92].

4.4 Permeability of Cells after Drying Rehydration

It was established that permeability of dried yeasts increases sharply during rehydration. The total losses of cell components may amount to 20–30% of dry weight [18]. Such an increase of cytoplasmic membrane permeability and a substantial leakage of intracellular substances explain the formation of bubbles, drops and clusters of various forms and sizes on the surface of rehydrated cells [18, 75].

A considerable leakage of free amino acids (Table 1) was observed after rehydration, whereas only traces of amino acids were released from intact cells. Considerable increase of the free amino acid pool in cells in a complex medium during reactivation

Table 1. Release of free amino acids from dehydrated *Saccharomyces cerevisiae* into water

Amino acid	Free amino acid content	Release of amino acids into water	
	mkg g^{-1}	mkg g^{-1}	% of content
Lysine	63.8	5.8	9.1
Hystidine	23.8	7.1	30.0
Arginine	33.2	3.4	10.4
Asparaginic acid	43.2	22.2	51.5
Methionine	3.0	3.1	100.0
Threonine	39.0	22.7	58.0
Serine	40.9	11.2	27.3
Glutamic acid	344.6	110.6	32.1
Proline	24.3	25.0	100.0
Glycine	7.6	8.0	100.0
Alanine	171.0	68.5	40.1
Cystine	47.8	14.1	29.5
Valine	61.5	17.1	27.8
Isoleucine	24.1	5.9	24.6
Leucine	21.8	4.8	22.0
Tyrosine	12.5	13.0	100.0
Phenylalanine	11.1	4.0	36.0

indicates the fact that the change of permeability accounts also for reverse processes — entering of substances into the cell. It was also shown that even a 1.5 h reactivation time decreases the permeability of their cytoplasmic membrane sharply and brings it close to that of the intact organisms [18,93].

It was noticeable, that dried yeasts lost about 50% of nicotinic acid and small amounts of thiamin and riboflavin after rehydration [94,95]. A considerable increase in dried cell permeability was found for compounds with peptide bonds with a typical absorption at 200 nm and for nucleotides absorbing at 260 nm (Fig. 14) [96].

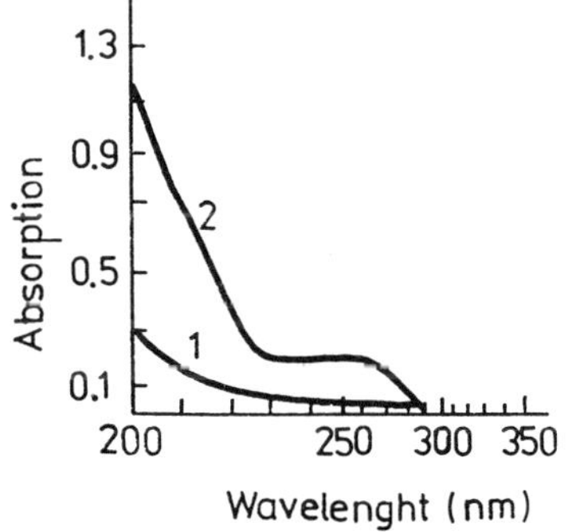

Fig. 14. Absorption spectra of yeast cell rehydration liquids.
1 — rehydration of initial cells;
2 — rehydration of dried cells

The release of potassium and magnesium ions from dried cells also notably increases (Fig. 13) [79,84,96]. It turned out that the amount of these ions excreted from the cells during rehydration does not correlate with the viability. An analogous absence of correlation between ion discharge and organism viability was detected when studying the effect of radiation on yeasts [97]. An increased release of protein compounds was observed, proportional to the decrease of viability. Electrophoretic investigation of proteins entering the rehydration medium, revealed a whole set of protein components, different as to their electrophoretic mobility and molecular weight [96,98].

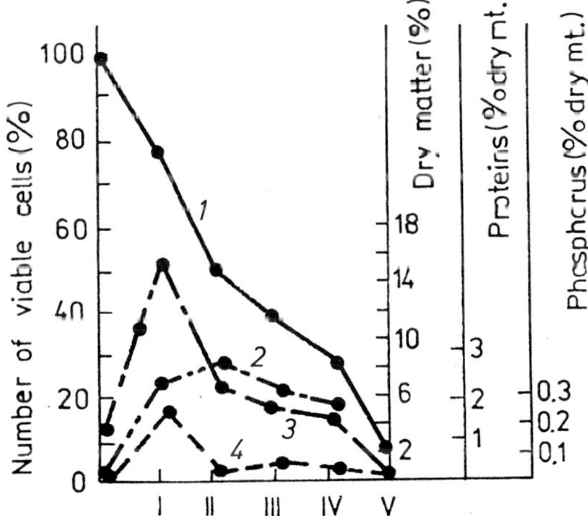

Fig. 15. Viability and permeability of yeasts during multiple dehydration-rehydration (I–V).
1 — viability; 2 — losses of dry matter; 3 — losses of protein substances; 4 — losses of phosphorus-containing substances

Changes of yeast cell populations after multiple drying and subsequent rehydration were studied experimentally. After the first drying about 25% of cells were inactivated, and 15% of dry matter leaked into the rehydration medium. After the fifth drying-rehydration cycle less than 10% of cells were viable and about 30% of dry matter were released into the medium, containing proteins, nucleotides, phospholipids, ergosterin and its esters (Fig. 15) [99, 100].

During investigations dealing with the increased permeability of dried yeast cells methods for studying the intracellular activity of enzymes have been worked out. Instead of cell homogeneization or treatment with dimethylsulphoxide, toluene, etc. a 2–3 h drying of cell suspension on filter paper at room temperature is now suggested [101].

All the data obtained testify to extensive functional and structural changes of the cytoplasmic membrane occurring during dehydration-rehydration.

Analogous experiments with dried bacterial cells have demonstrated that substances leaking into the rehydration medium can be assimilated by plants and may positively affect their growth after external environment stress [18].

4.5 Changes of Intracellular Yeast Membranes by Drying-Rehydration

At present the most adequate membrane model is proposed by Singer-Nicolson. It is supposed that proteins and lipids are capable of a lateral heat diffusion in the plane of membrane [102]. Membrane contains small-sized lipid domains, which are characterized by various physico-chemical characteristics of their lipid components [103]. There exist data, confirming the presence of close protein-protein interactions in the membrane and formation of a network of protein aggregates on some parts of the membranes [104, 105]. Thus, for yeasts heterogenous cytoplasmic membrane fragments of different floating density and various enzyme and lipid concentrations have been demonstrated [106]. It is supposed that there exists a connection between this structural heterogeneity of membrane and the functional spatial compartmentalization and intergration [107]. It was detected, that many membrane proteins and enzymes are separated from the lipid phase by a layer (or layers) of the so-called "boundary lipid" whose rotational and lateral diffusion decreases. In contrast the fatty acid tails in lipids of this layer are of low regularity [108, 109]. The presence of this layer is supposed to foster a decrease in enzyme sensitivity to changes in the lipid matrix of the membrane [110]. It is assumed that the state of the system "boundary lipid — total lipid phase" may determine the regulation of protein aggregation in membranes [111, 112]. Membrane enzymes, hence, may prove to be sensitive to changes in the microviscosity of lipid environment, if there are any essential interrelations between their subunits and molecule motion upon their functioning [107, 113]. Water plays a special role in biological membranes. The very appearance of bimolecular lipid layer with strictly oriented phospholipid molecules is possible only in a water medium due to hydrophobic interactions between molecules. Water makes up 30–50% of membranes and 20–30% of membrane water is bound [42]. These data are in keeping with the results of studies, which demonstrated that upon hydration of dried organisms membrane systems restore their normal configuration upon a 20–30% relative moisture of the cells [114]. Prior to this it was found that just 20–30% of water is sufficient to preserve the phospholipid bilayer in membranes [115,116]. Such amount of water maintains membrane lipids in lamellar phase. A further dehydration may cause a molecular reorganization of lipids in membrane [117].

As it was already noted, yeasts pass into the state of anabiosis at 8–10% moisture content, i.e., when all the free and a part of bound water have been extracted. Based on the above considerations, in 1977 we suggested a model of transformations in separate membrane parts during cell dehydration. It is supposed that the loss of water hydrate layer causes disorientation of phospholipid molecules, at least in some membrane parts (Fig. 16). Under these conditions membrane lipids may change

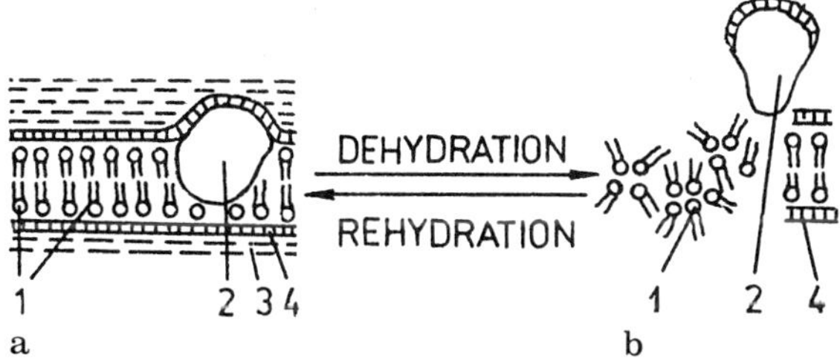

Fig. 16a and b. A hypothetic model of transformations in a part of biomembrane during dehydration and rehydration **a** native membrane; **b** a part of membrane after dehydration. 1 — lipid molecules; 2 — membrane protein; 3 — free water; 4 — bound water

their configuration from lamellar to hexagonal —, when their orientation is determined by electrostatic powers [118–119]. An analogous conclusion was made later by Simon, too [120]. Experimental proof of the possible appearance of such transformations under real conditions of biological object dehydration was obtained in model experiments on vesicules of sarcoplasmic reticulum, as well as in studies on nematodes *Aphelencus avenae* and on the cells of rye leaves [121,122], while studies on desiccation-tolerant moss *Tortula ruralis* did not result in an analogous conclusion [123]. Yet, in the latter case the moisture content of the object of desiccation was, most probably, "insufficient", making up only 18%, while in the rest of the above works the moisture contents of membranes under study were much lower. Such disorientation of phospholipids in membranes of dried organisms increases the danger of their irreversible damage under these conditions. This may be due to oxygen leakage and the presence of oxidizing reactions, with peroxidation of lipids and with the presence of phospholipases and enzymatic degradation of the main biomembrane components. A consequence of membrane phospholipid disorientation is the destroyal of biomembrane barrier function, testified by the increased permeability noted above.

5 Biochemical Changes in Yeast Cells During Drying, Rehydration and Reactivation

5.1 Changes of Nucleic Acids

During drying of microorganisms the bulk of intracellular compounds undergo certain changes of multivariate character: some of them are aimed at preserving cell viability under the extreme conditions of dehydration, the other may cause a partial inactivation of population under these conditions, and still other may have practically no effect on the state of dried organisms.

Nucleic acids are one of the basic intracellular compounds, being of interest in the studies on the effect of various extreme conditions on the cell. Changes of water content are known to affect the structural organization of nucleic acid molecules. Water, surrounding DNA molecule, enhances the stability of its conformation [124], and dehydration may bring about a collapse of the structure of macromolecules [125, 126], and cause anomalous biological properties of DNA [127]. Drying may cause also the rupture of DNA strands [128]. The available data indicate to the fact, that under these conditions there may be essential changes in some of DNA biophysical parameters, too [129].

The above mentioned DNA changes were established when studying the effect of freeze-drying on isolated purified macromolecules. At the same time, when studying the anabiosis of living organisms, information on DNA changes in vivo is more important. Such data on yeast organisms, unfortunately, are scarce as yet.

When studying the effect of convective drying a condensation of chromatine was observed in cells (see above), which can be regarded as one of the adaptive reactions in cells, aimed at protecting DNA from damages and preserving organism viability [61, 64, 65]. At the same time it was established that the resistance of DNA to weak alkaline hydrolysis may notably decrease during drying of cells [59].

The process of drying does not cause any serious genetic changes in DNA. At the same time, there are quite serious changes, tested as to the appearance of adenine-dependent mutants, which are due to the effect of oxygen on dried yeast cells [130, 131].

Analysis of the quantitative changes of nucleic acids demonstrated that after dehydration of yeast cells the number of nucleic acids diminished in the alkali-soluble ("labile") fraction and practically did not decrease in the hot perchloric acid extract ("stable" fraction). A partial destruction of nucleic acids and, most probably, of other nucleotide-containing compounds causes an increase in the amount of nucleotides (acid-soluble fraction) (Table 2) [132].

Closely related *Saccharomycodes ludwigii* strains, different in their resistance to freeze-drying, were compared as to their nucleic acid content. The results demonstrated that the resistant strains have a higher nucleic acid content in the "stable" fraction (Table 3) [132]. It was concluded, that a lower nucleic acid content in this fraction is one of the factors, determining the lability of yeasts by drying.

Another work demonstrated that during convective drying up to 40% of RNA's

Table 2. Nucleic acid content in intact and dried *Saccharomyces cerevisiae* cells

Cells	Fractions					
	Acid-soluble		Alkali-soluble		Hot perchloric acid extract	
	μg g^{-1} dry wt	%	μg g^{-1} dry wt	%	μg g^{-1} dry wt	%
Intact	617.2	100	322.9	100	210.0	100
Dried	1060.0	172	212.4	66	208.0	99

Table 3. Content of nucleic acids in intact *Saccharomycodes ludwigii* cells

Strains	Characteristics	Resistance to dehydration	Nucleic acid content			
			"Labile" fraction		"Stable" fraction	
			μg g^{-1} dry wt	%	μg g^{-1} dry wt	%
Saccharomycodes ludwigii BKM Y-630	Initial strain	Nonresistant	2529.3	83.6	495.8	16.4
Saccharomycodes ludwigii BKM Y-631	Mutant strain	Resistant	1007.1	74.0	353.6	26.0
Saccharomycodes ludwigii BKM Y-638	Initial strain	Resistant	1255.0	79.3	334.0	20.7
Saccharomycodes ludwigii BKM Y-639	Obtained from one spore of a 4-spore strain *S. ludwigii BKM Y-638*	Nonresistant	1948.5	87.2	286.7	12.8

may be destroyed. RNA degradation takes place mainly at the first stages of drying, when moisture content decreases to about 20%. It was suggested that RNA degradation may be linked with enzymatic processes [18, 133].

Data were obtained which demonstrated a simultaneous synthesis of both, RNA and protein immediately at the beginning of reactivation of dried yeast cells. This indicates the fact that in the process of convective drying there remain in dried cells all kinds of RNA in an undamaged state, in amounts sufficient to initiate protein synthesis [18].

5.2 Effect of Drying on Protein Compounds

Biochemical investigations revealed certain changes of proteins, taking place during drying of microorganisms. The proteolytic enzyme activity [134] and the possibility of proteolytic processes taking place in cells [135] were observed to increase under these conditions. While during convective drying, which preserved a considerable number of cells viable, only a partial and selective proteolysis was observed [136].

Studies of protein fractional composition in dried *Saccharomyces cerevisiae* demonstrated a different resistance of various protein fractions to drying, this resistance being dependent on growth conditions, too. Proteins, dissolved in tris-buffer (Table 4), were found to be the most sensitive to drying, and the greatest decrease of protein content in this fraction after drying was observed in yeasts grown on media with an increased content of molasses and ammonium salts, containing, consecutively, 9% of total nitrogen. It was established, that cells with

Table 4. Content (in % from total N × 6.25) of protein fractions in yeast *Saccharomyces cerevisiae*, depending on drying conditions

Drying conditions	Protein fractions, to be extracted by			
	Tris-buffer	5% NaCl solution	70% ethanol solution	0.2% NaOH solution
Initial yeasts	46.4 ± 5.0	5.2 ± 0.2	0.75 ± 0.05	7.3 ± 0.5
Freeze-drying	39.0 ± 5.0	4.1 ± 0.5	0.82 ± 0.08	11.3 ± 0.6
Drying at 35 °C, 2 h	30.9 ± 3.8	5.5 ± 0.4	0.90 ± 0.07	9.8 ± 0.4
Drying at 70 °C, 20 min	25.3 ± 4.5	3.2 ± 0.3	0.75 ± 0.05	11.5 ± 0.3
Drying at 70 °C, 24 h	21.0 ± 4.5	3.1 ± 0.2	0.60 ± 0.05	9.2 ± 0.2
Drying at 105 °C, 20 min	19.8 ± 3.5	2.9 ± 0.1	0.70 ± 0.05	9.1 ± 0.4
Drying at 105 °C, 24 h	13.0 ± 2.6	1.6 ± 0.2	0.46 ± 0.05	3.0 ± 0.2
Drying at 140 °C, 15 min	9.0 ± 2.8	1.0 ± 0.2	0.28 ± 0.05	2.1 ± 0.2

total nitrogen content of 9% after freeze-drying contain 36% of the initial protein amount in this fraction, while in yeasts with 6.2% of nitrogen — 62% [53, 137]. Under these conditions proteins, soluble in a 5% sodium chloride, were slightly more stable. Protein fractions, soluble in a 70% ethanol and a 0.2% NaOH solution were even more resistant to drying.

Since the experiments did not detect any notable changes in the protein amount in cells during drying, the observed effects are not due to protein destruction, but due to the changes of its physiochemical properties, solubility in particular [18].

Studies on the effect of drying on the electrophoretic spectrum and protein content in the tris-buffer fraction revealed a re-distribution of proteins after drying — "disappearance" of some gel zones, peculiar to intact yeasts, and appearing of some new ones [18]. After convective drying of yeasts, containing 9% of nitrogen (DM), 7 components (out of 18) changed their electrophoretic mobility, while in yeasts containing 6.2% of nitrogen — only two components. Freeze-drying caused even more essential changes in the protein spectrum of the tris-buffer fraction of yeasts [138].

Changes in the electrophoretic spectrum and protein content in the salt-soluble fraction under various conditions of drying were negligible. Some changes were detected in the relative electrophoretic mobility of the most "mobile" proteins. After drying the sum of components in this zone was the same as in initial yeasts, but they contained a little less protein. Yeasts with an increased nitrogen content had a shift in the electrophoretic mobility of protein components towards most mobile proteins [18, 138].

Studies on the changes of various enzyme activity in dried yeast cells established that they may greatly depend on growth conditions and drying methods. These investigations detected, that some hydrogenases tend to increase their activity in the results of drying (Table 5). The greatest increase of their activity was detected in yeasts grown on a sugar-rich medium. Dehydrogenase activity increased due to an increase of its isosyme activity [139].

Phosphohydrolase activities in yeast cells were studied. It was established that after freeze-drying the activity of adenosine triphosphatase and pyrophosphatase decrease

Table 5. Effect of drying on some dehydrogenase activities in yeasts *Saccharomyces cerevisiae*

Carbon source in nutrient medium	Total nitrogen content in yeast cells	State of yeast cells	Enzymes							
			Alcohol dehydrogenase		Malate dehydrogenase		Glucoso-6-phosphate dehydrogenase		NADP-Dependent glutamate dehydrogenase	
			Number of isosymes	Total activity of enzymes	Number of isosymes	Total activity of enzymes	Number of isosymes	Total activity of enzymes	Number of isosymes	Total activity of enzymes
Molasses (8% sugar)	9.0 ± 0.2	Intact	[illegible]	97.5	4	78.2	3	51.8	1	41.0
		Dried	[illegible]	150.5	4	174.0	3	72.8	1	58.5
Molasses (4% sugar)	7.5 ± 0.3	Intact	3	56.0	7	236.0	4	185.5	1	82.5
		Dried	[illegible]	13.3	6	196.0	5	295.8	1	53.5
Ethanol (1.5%)	8.3 ± 0.1	Intact	[illegible]	125.0	6	233.7	2	65.5	1	64.0
		Dried	4	114.5	6	276.2	2	52.8	1	44.0

Table 6. Specific activity of phosphohydrolases (mU mg^{-1} of protein per min) in yeast cells in function of cultivation and drying conditions

State of yeast cells	Polyphosphatase		Tripolyphosphatase		Pyrophosphatase		ATP-ase	
	1	2	1	2	1	2	1	2
Intact	63	115	39	62	2830	1690	84	66
Dried at 37 °C for 24 h	54	101	37	55	2160	1850	63	60
Freeze-dried	49	94	31	56	1860	ND	56	37

Note: Biomass was grown on: 1 — molasses medium;
2 — on Reader's medium, containing 0.2% of yeast extract

to some extent. There was observed also other, though not very notable, decrease (both on convective and freeze-drying) in polyphosphatase, tripolyphosphatase and pyrophosphatase activities (Table 6) [140].

The activity of alkaline phosphatase in either convectively or freeze-dried cells practically did not change, while a sharp (by 2–3 times) increase of acid phosphatase was observed in comparatively slowly dried yeast cells. It was concluded that the activity of acid phosphatase increases because it is being synthesized *de novo* during drying of cells [141–143].

There was observed some increase in cytochrome-c oxidase during drying of cells grown on molasses, and a slight decrease in those grown on a synthetic ethanol medium [18]. No changes of isocitrate lyase activity could be traced [144], and practically that of invertase, too [145]. There are data on an increase of catalase, peroxidase and carboxylase activities during drying and freeze-drying in particular [146].

It is supposed that such changes in enzyme activity, taking place during drying of yeast cells, besides synthesis of enzymes *de novo* may be connected with changes in protein molecule conformation. Since all the free and a great part of the bound water have been extracted from cells during drying, it may cause a "destroyal" of the old ones and "appearance" of new active centers in enzyme molecules [18]. This hypothesis is supported by the fact that enzyme activities in protein-enriched cells change more notably. This may be due to a greater bound water content. When drying such organisms to the usual moisture content, this is accompanied, obviously, by greater amounts of bound water extracted from the cells and further on — by more notable conformational changes in protein molecules.

5.3 Changes in the Composition of Lipids in Cells

Lipids and their fatty acid composition are known to determine the expression of the main functional properties of intracellular membranes [147–149]. The main biochemical mechanisms of lipid accumulation in yeast cells have been studied in great detail [150–156]. Fatty acid composition in membrane lipids affects the activity of yeast cytoplasmic membrane transport proteins. It is supposed to be connected with the changes in membrane lipid fluidity by varying the amount of unsaturated fatty acids [157]. Depending on whether cytoplasmic membrane phospholipids contain more or less phosphatidylethanolamine and phosphatidylcholine, the sensitivity of yeast cells to surface active compounds, osmotic shock, increased temperatures and other extreme effects changes [158–163]. There have been demonstrated the possibilities of a directed modification of fatty acid composition of yeast lipids by varying the conditions of cell cultivation and depending on the stage of cell cycle [164,165].

Studies on the changes of lipid molecules, appearing during yeast drying, are comparatively scarce as yet. Drying nitrogen-rich yeasts convectively revealed a diminishing of phospholipid amount in cells by about 10–20%, mainly lecithin and phosphatidylethanolamine. There was also detected an increase in the amount of free fatty acids in cells. The obtained results allowed to conclude, that the observed effects are due to phospholipase [51]. It was established, that phospholipase activity in yeast cells may increase almost two-fold during drying [166]. Yet a series of experiments on the effect of drying on *Saccharomyces cerevisiae* did not detect any notable quantitative changes of total lipids and qualitative changes in their fractional composition. It was demonstrated that after drying the lipids contain more short-chain fatty acids (C_{10-14}) and a redistribution of palmytic and palmytooleic acid content can be observed in them. Analogous phenomena were observed also in the phospholipid fraction [167–168]. It was established, that under these conditions during dehydration-rehydration there is an increase in the ratio between the unsaturated and saturated fatty acids (unsaturation degree) in total lipids and phospholipids in yeasts grown both, on rich molasses and on a synthetic ethanol medium. The most notable changes in these fatty acid ratios were observed in yeasts, less resistant to drying and rehydration. Fatty acid unsaturation degree in resistant intact cells was lower than the corresponding one for more sensitive to drying intact organisms [168].

Experimentally it was established that yeasts, taken from the logarithmic growth phase at the maximum of fatty acid unsaturation degree, were the least resistant to drying. Addition of glutamic acid or biotin into a synthetic medium, which increases yeast resistance to drying, diminishes the fatty acid unsaturation degree [168].

The established reverse dependence between the fatty acid unsaturation degree of yeast lipids and their viability after drying was found to be valid only to a certain extent, corresponding to a 30% viability [168].

Mechanisms of the changes in the unsaturated and saturated fatty acid ratio during drying towards an increase of the former have not yet been investigated in detail. This is supposed to be connected with the fatty acid desaturase activation and itself is one of the adaptive reactions of the cells, aimed at the preservation of their viability (in the given case, at maintaining the fluidity of biological membranes during cell drying) [168].

A comparison was made of lipid fatty acid unsaturation degree, alcohol dehydrogenase activities and trehalose content during drying and rehydration of yeast cells. The distribution of the initial levels and character of changes in trehalose content and alcohol dehydrogenase activities were found to be similar to the changes of fatty acid unsaturation degree. It was concluded that there was a strong correlation between fatty acid unsaturation degree in dried organisms and alcohol dehydrogenase activity changes [169,170].

Changes in the lipid fatty acid unsaturation degree, taking place upon drying-rehydration, are interrelated with the respective changes of alcohol dehydrogenase and glucose-6-phosphate dehydrogenase.

In yeast dehydration one of the factors affecting the indices of fatty acid desaturation degree in intact and dried yeasts, is the activity of mitochondrial ATP-ase. Populations more resistant to drying-rehydration can (as compared to the ones sensitive to the given effect) change the relation between the activity of mitochondrial ATP-ase activity, alcohol dehydrogenase and glucose-6-phosphate dehydrogenase more notably mainly by increasing the latter [171].

The effect of storage on the lipid content in dried yeast cells was investigated. During storage the number of free fatty acids in dried yeast cells increases. The most intensive changes of this type could be observed in dried yeasts, stored at higher temperatures [172].

A further development of these investigations will, no doubt, cast more light on the changes in the lipid composition, taking place during drying.

5.4 Changes of the Carbohydrate Content in Cells

The total amount of carbohydrates in intact yeast cells depends on the cultivation conditions, growth phase and some other factors, and it may reach 50% of cell dry matter [173].

When studying the effect of drying on yeast carbohydrates, it was established that the basic cell wall components — mannane and glucane — practically do not change under these conditions [174], though structural and functional changes were detected in them. A certain decrease of glycogen during *Saccharomyces cerevisiae* drying was observed [175]. Yet, when studying the carbohydrates, the main attention is paid to trehalose content in cells. It is well known that organisms, capable of anabiosis by drying, accumulate great amounts (10–15%) of trehalose [176, 177], and that the quality of dried yeasts is in direct dependence on trehalose content in cells [178]. The importance of trehalose for organisms capable of anabiosis is due to the fact, that it does not take part in reactions between the reduced groups of sugars and free amino acid groups of dry proteins [4], even more — it inhibits these reactions, forming a shell round the proteins [179]. Trehalose is supposed to be able to stabilize membranes, too [180]. Experiments have demonstrated, that like a soft drying and sporulation of *Saccharomyces cerevisiae* is accompanied by an increased trehalose content [181, 182]. Also an intensive drying at 105 °C for example, causes a notable decrease of trehalose [175]. Reactivation of dried yeasts at optimum conditions causes a rapid decrease of trehalose content [183], like spore germination [184] and cultivation of intact cells [185]. Yet it turned out that a certain amount of trehalose may leak into the medium during rehydration. Thus, even in yeasts with a very high ratio of viability (93%), about 16% of trehalose leaked in 10 min [183]. Towards the middle of the reactivation process up to 60% of the leaked trehalose was observed to be reabsorbed by the cells again [183]. The specific activity of trehalose either during drying or subsequent rehydration remained practically unchanged [183]. This in contrast to spore germination, when the enzyme was observed to become activated [184]. Hence it was assumed that trehalose is not used as a carbon and energy source during rehydration in its early stages [183].

5.5 Polyphosphate Changes in Cells

Of late great interest has been paid to polyphosphates [186]. Studies have been conducted concerning the changes of inorganic polyphosphate content of *Saccharomyces cerevisiae* cells, as well as the activity of a number of enzymes of their metabolism under various drying conditions and degrees of dried organism viability. Growth conditions were found to affect the total accumulation of polyphosphates greatly. It was established that the total amount of polyphosphates was not affected by drying. Yet, an essential redistribution of various polyphosphate fractions was detected to take place during drying and subsequent rehydration (Table 7). All the studied methods of cell drying increased the number of acid-soluble polyphosphates and decreased (with an exception of freeze-drying with a protective medium) their content in salt-soluble polyphosphate fraction. Reactivation of dried yeasts decreased the amount of acid-soluble and increased that of salt-soluble polyphosphates [140]. Cytochemical electron microscopy detected that after yeast dehydration there takes place a redistribution of polyphosphate localization in the cell. There was observed also an increase of the number of polyphosphate-containing bodies in the cytoplasm and cell wall and a decrease of their number in vacuoles [187]. Under various conditions of drying and subsequent reactivation a direct correlation was established between the salt-soluble polyphosphate fraction and the total sum of nucleic acids. A reverse dependence was demonstrated — that between the amount of salt-soluble polyphosphates and nucleic acids on the one hand, and the amount of acid-soluble

Table 7. Content of phosphorus compounds (P = μg g^{-1}) in *Saccharomyces cerevisiae* biomass

Compound	Biomass grown on molasses		Biomass grown on Reader's medium with 0.2% yeast extract		
	Initial	Dried at 37 °C for 24 h	Initial	Dried at 37 °C for 24 h	Freeze-dried
High-molecular polyphosphates (total)	3010	4930	8990	10270	10560
Acid-soluble polyphosphates	1510	1760	4250	5820	6050
Salt-soluble polyphosphates	1290	580	2080	1210	1690
Alkali-soluble polyphosphates	—	—	1840	2690	2690
Polyphosphates, soluble in hot $HClO_4$	2210	2610	820	550	130
Ortho-phosphate	1390	2370	3580	3420	3060
Nucleic acids	5890	5300	—	—	—
Nucleotides	760	1260	520	960	820

polyphosphates — on the other [140]. It was demonstrated that there exists a relationship between the indices of yeast cell viability and both, trehalose content in them and the content of high-molecular polyphosphate fractions. It is supposed that relationship between cell viability and trehalose content is determined by the level and changes of two fractions of high-molecular polyphosphates [188].

No essential changes were detected in ATP-ase, pyrophosphatase, polyphosphatase and tripolyphosphatase activities in convectively dried cells as compared to the initial culture, disregarding the extremely low viability of dried organisms in some experiments (Table 6). Some decrease in the activities of phosphohydrolases under study was observed only during the freeze-drying of yeasts. It was concluded that the redistribution of polyphosphates into different fractions during drying and rehydration of yeasts takes place not due to the effect of phosphohydrolases under study. The redistribution of high-molecular polyphosphate content into different fractions can be due to the changes in the nucleic acid metabolism. Phosphorus from nucleoside monophosphates, produced by nucleic acid degradation, is supposed to be transported to polyphosphates [140].

There has been established a reverse dependence between cell viability upon drying and the content of polyphosphates [140].

6 Restoration of Yeast Populations

6.1 Rehydration

As it was shown above, the processes of drying and rehydration essentially affect the structural organization and metabolism of microbial cells. It is obvious, that not only drying, but also rehydration of dried cells may prove to be a detrimental factor. At times it is quite difficult to distinguish between these two processes — which of them has caused the changes.

When reactivating microbial population from anabiotic state, three processes can be distinguished:

- moistening of the preparation (rehydration),
- reparation of the damaged cell structures and macromolecules, and re-synthesis of a great variety of compounds, to replace those degraded during dryingrehydration processes (reactivation),
- restoration of the former number of population by multiplication of viable cells (cell growth and multiplication).

Rehydration (moistening) is a purely physical process, when cell saturation with water takes place. Though the kinetics of dried cell moistening with water is poorly studied as yet, it is known, that the time required for the return of water into cells is not long. One may assume that during rehydration a part of reversibly damaged cell elements obtain their initial native expression (e.g., the relations "water-protein", "water-nucleic acids", "water-phospholipids", etc. are formed anew).

Rehydration of dried granulated baker's yeast is over in several minutes, if the

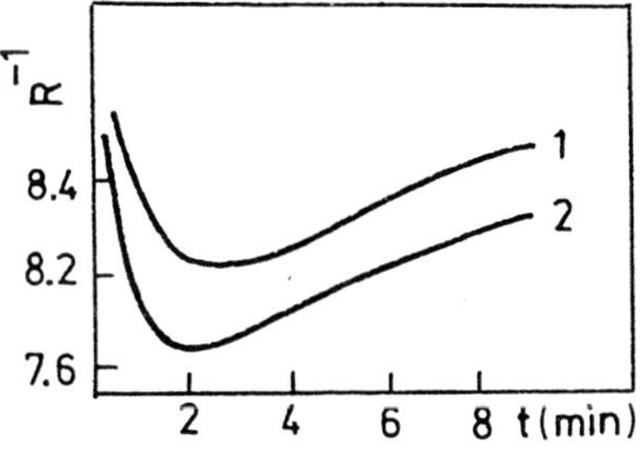

Fig. 17. Electric conductivity of dried yeasts upon their hydration.
R^{-1} — electric conductivity;
1–20 kHz; 2–4 kHz

size of the granules does not exceed 1.0–1.5 mm. It is testified by the changes in electric conductivity (Fig. 17). Right after rehydration the electric conductivity is high, that may be due to a dense layer of extracellular water around the granules containing cellular electrolytes. In 1–2 min this extracellular water is absorbed by the cells but all the spaces in cells and the midst of granules is filled up with water in 8–10 min. During this period of time the electric conductivity increases up to the initial level [79].

It is advisable to start the rehydration of dried cells gradually, in atmosphere saturated with water vapour, at 0.01–0.1 mg of water per 1 mg dry weight of biomass per s [189]. It proved true on the model of dried baker's yeast [44]. The temperature of the medium is of importance, too. Losses of intracellular components decrease if the rehydration is carried out at higher temperatures. Thus, 30 °C is the optimum temperature for yeast growth, but 43 °C — for rehydration of dried cells. Upon 4–5 °C during rehydration, yeasts lose 75% of diphosphopyridin nucleotides, upon 43 °C — only 15% [190]. Leder [191] considers that low temperatures cause crystallization of membrane lipids and in membranes there appear channels.

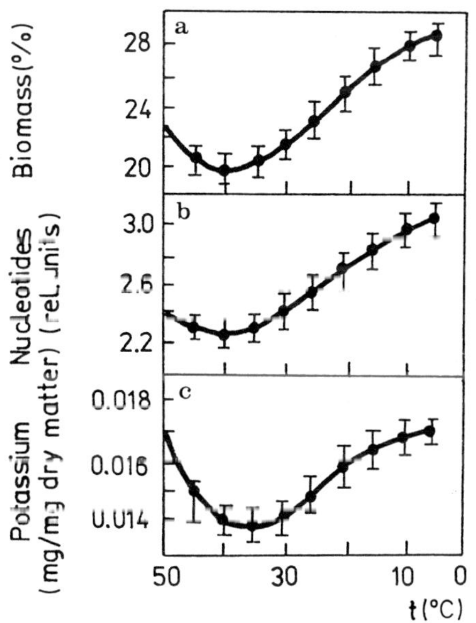

Fig. 18a–c. Dependence of total intracellular substance losses (**a**), nucleotide (absorption of rehydration liquid at 260 nm) (**b**) and potassium (**c**) leakage on the temperature of dried yeast rehydration

Changes of permeability is one of the most characteristic criteria of damages in the cytoplasmic membrane under effect of any detrimental factors. It was established that the minimum losses of dry matter, nucleotides and potassium ions during rehydration of dry yeast cells are observed at 35–45 °C (Fig. 18) [192]. The positive effect of increased (the optimum for the growth of the given culture is 30 °C) temperatures of rehydration can, obviously, be explained by improved conditions for phospholipid hydration, since the latter is difficult at temperature lower than that of phase-transition, but in "dried" phospholipids the temperature of phase-transition is increased [193–195].

There exist data that the amount of leakage of cell components during rehydration depends on the composition of the medium [196–198]. In dried yeasts to be rehydrated the maximum permeability of cytoplasmic membrane for nucleotides was observed upon a 3–5% concentration of NaCl or KCl in the medium [192]. Since monomolar solutions of sodium chloride are known to cause a re-orientation of phospholipid polar head groups [199,200], it is supposed that in the given case in solutions with similar salt concentrations there is an increase in the disorientation of lipid components in biomembranes caused by the removal of a part of bound water at the stage of dehydration.

Table 8. Relative losses of nucleotides during rehydration of dried yeasts depending on glucose and calcium content in the medium.
(Nucleotide losses in the absence of glucose and calcium are assumed to be 100)

Ca ($mmol\ l^{-1}$)	Glucose ($mmol\ l^{-1}$)		
	0	30	150
0	100	93	81
3	92	81	70

Calcium ions and glucose, too, demonstrate a stabilizing effect on membranes at the rehydration (Table 8) [192]. Binding themselves to the polar head groups of two neighbouring phospholipids, calcium ions form salt bridges, thus increasing the rigidity of membrane structure [201]. Glucose may permeate into the cell and stimulate the formation of protein gels, that prevents component diffusion from the cell. On the other hand, at higher temperatures glucose may accelerate sugar-amine reactions and thus increase cell permeability. Calcium ions may diminish this negative effect of glucose [202].

6.2 Reactivation of Yeasts

After the rehydration of cells the resources of energetic and constructive metabolism and the complex of the optimum conditions (for the given organisms) being present, a reactivation of reversibly changed organisms sets in. The duration of reactivation

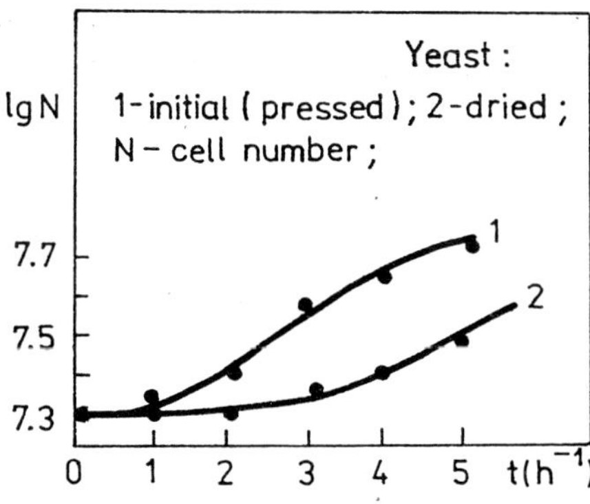

Fig. 19. Kinetics of compressed yeast cultivation and dried yeast reactivation

period depends on the character and level of intracellular damages. It is considered [203] that the following processes may take place upon reactivation:

- biosynthesis of new molecules (macromolecules included) instead of the damaged ones;
- compensation for the function of the damaged structures with the help of reserve mechanisms;
- destruction and isolation of partly damaged structures;
- reparation of reversibly altered structures.

It was established [184] that dried yeasts inoculated in a nutrient medium have a longer lag-phase (by 1–2 h) as compared to the initial compressed yeast (Fig. 19). Depending on the character and level of damages, the specific growth rate μ_0 of dried yeast cells may be lower than that of the initial compressed ones. In order to characterize the level of cell damages and the adaptability of yeasts to the medium of reactivation, there have been worked out a mathematical model and introduced the value M_0 [18]:

$$M_0 = \frac{\mu_0}{\mu_{max}},$$

where μ_{max} — the maximum specific growth rate.

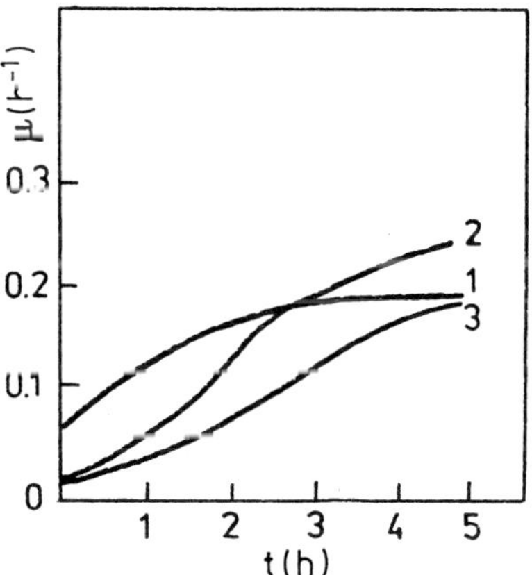

Fig. 20. Changes in the specific growth rate of compressed (1), dried (2) and freeze-dried (3) yeasts on a semisynthetic medium with glucose. The curve was obtained by calculations according to the model [18]

Both experimentally and by calculations it was established that, irrespective of the medium composition and the method of drying, during reactivation of dried yeasts the specific growth rate is almost fully restored to the value, characteristic of the initial compressed yeasts.

Figure 20 shows the calculation dynamics of the specific growth rate of initial, convectively- and freeze-dried yeast samples during their subsequent cultivation on a semi-synthetic medium with glucose in the presence of biotin and pantothenate as to the above model [18]. Characteristics of yeast growth are presented in Table 9. M_0 was determined by way of calculations from experimental data [18]. It was established that upon drying and freeze-drying M_0 decreases more than 10 times.

During rehydration the specific oxygen consumption rate is sufficiently rapidly fully restored in dried and freeze-dried yeasts. It rapidly increases starting already with the first minutes of reactivation. Supposedly, during this period the cells undergo the processes of reparation, which require energy expenses [197].

Table 9. Specific growth rate of native, convectively dried and freeze-dried yeasts during subsequent cultivation in a semi-synthetic medium with glucose

Yeasts	M_0	μ_{max}, h^{-1}	τ, h
Native	0.26	0.21	0.9
Dried	0.015	0.24	0.84
Freeze-dried	0.017	0.21	0.84

6.3 Restoration of the Growing Population

Since a part of cells becomes irreversibly inactivated when dried, after reactivation a certain time is needed for the multiplication of the viable individuals and restoration of the number of population. This, obviously, requires a rich nutrient medium and the optimum regimes of cultivation. After reactivation, which in this case coincides with the prolonged lag-phase, the exponential growth phase sets in, and the growth of biomass can be expressed as follows:

$$X = X_0 \, e^{\mu(t_1 - t_0)} ,$$

where: X — the initial viable biomass prior to drying,
X_0 — viable cell biomass after drying, storage, rehydration, reactivation,
e — basis of the natural logarithm (2, 718, ...),
μ — specific growth rate for the given culture, h^{-1},
$t_1 - t_0$ — time of cultivation, h.

In the given case $t_1 - t_0 = \tau_3$,
where τ_3 — time required for the multiplication of the viable individuals and restoration of the number of population.

To establish the time of multiplication, the equation was logarithmized and the following was obtained:

$$\tau_3 = \frac{\lg X - \lg X_0}{0.43\mu}.$$

If the viability of population (I_S) were determined after reactivation, there can be obtained:

$$I_S = \frac{X_0}{X} \quad \text{and} \quad X_0 = I_S X.$$

Introducing X_0 in the given equation —

$$\tau_3 = \frac{\lg X - \lg I_S X}{0.43\mu} = \frac{\lg X - \lg I_S - \lg X}{0.43\mu},$$

$$\text{i.e., } \tau_3 = -\frac{\lg I_S}{0.43\mu}.$$

Time of multiplication is seen to be proportional to viability and inversely proportional to the specific growth rate. μ is known to be established by Monod's equation:

$$\mu = \frac{\mu_{max} S}{S + K_m},$$

where S and K_m are known from Michaelis-Menten equation; μ_{max} — the maximum growth rate, when substrate S is not a limiting factor.
Time of reactivation S assumed to be a constant value.

$$\tau_3 = -\frac{\lg I_S}{0.43\left(\frac{\mu_{max} S}{S + K_m}\right)} = -\frac{S + K_m \lg I_S}{0.43\mu_{max} S},$$

where: S — substrate concentration,
K_m Michaelis' constant,
I_S — viability of population,
μ_{max} — the maximum specific growth rate, when the substrate is not a limiting factor.

Consequently, the time required for restoring the number of population depends on viability, specific growth rate, substrate concentration and Michaelis' constant. The minus means that viability is always lower than 1.

Thus, a number of complicated biochemical processes takes place in dried microorganisms after dehydration, which fully restore the initial physiological status of the population.

7 Application of Dehydration for Production of Active Dry Yeasts

The main requirement, placed upon the technology of microorganism dehydration, is the preservation of them viability. For this purpose, a selection of strains, resistant to drying, is being carried out, microorganisms are grown under conditions ensuring the formation of resistant cells and sparing methods of drying are used.

It has been established that resistance of yeasts to drying varies greatly within various strains of the same species [204,205]. In the USSR, for example, in order to obtain active dry yeast, either strain *Saccharomyces cerevisiae 14*, or a special strain *Hp-2-Uzlovskaya* are used [56].

The medium for yeast growth must be rich, yet there must not be any extensive accumulation of nitrogen in the biomass (the optimum is 6.5–7.5% d.m.). An increased content (up to 10–20%) of trehalose is compulsory [56,205,206]. Trehalose formation is stimulated by an increased (37 °C) temperature and diminished aeration at the final stage of yeast cultivation. Trehalose accumulation in cells in enhanced also by 1 h of starvation 2–3 h prior to the end of growth. Some factories practice the cultivation of yeasts to be dried on media containing the salts of alkaline or alkaline-earth metals [56]. A method for obtaining a variety of yeasts resistant to drying, envisages their growth on a sugar-deficient medium, with a rapid subsequent cooling of the biomass by aeration, and an addition of 0.1–3% of sugar [207]. Yeast resistance to dehydration can also be improved by growing them in the presence of thiamine chlorohydrate, added in 0.05–5 mg portions per 1 L of culture medium [208].

After growth yeast is separated from the medium and compressed. The cake with 72–76% moisture is usually granulated. In the practice of yeast production yeast granules are dried convectively on driers of various design, upon regimes not exceeding 30–35 °C in the biomass.

In the 40ies–50ies in the USSR drying chambers were used most widely, as well as driers VIS-42D [49]. To intensify the process of drying, yeast granules were suspended and dried on air-fountain type and fluidized bed driers. This allowed to increase the temperature of drying agent up to 60–80 °C at the inlet, without danger of overheating the cells. The duration of drying in this case was thus shortened to 20–60 min [49]. Some yeast factories in the USSR at present use an industrial scale vibrodryer Al-VGS with an 80–100 kg h^{-1} output of dried yeast. It has four beds in a row; air temperature —50 °C, the whole process lasts 2.5–3.0 h. In order to ensure the desirable moisture of yeasts it is advisable to use air conditioners in the last two sections of the drier [56].

In many countries of the world, in a commercial production of active dried yeasts, drum-type driers are used, and several other methods of dehydration have been suggested. For example, a vacuum drying of yeasts upon a 50 mm Hg residual pressure and a decreasing temperature (40–80 °C at the first stage of the process, 35–65 °C at the second and 30–55 °C at the third one) [209,210], as well as various modification of yeast drying in a fluidized bed [211–214]. It is possible to increase yeast temperature over 50 °C towards the end of drying, when dry matter content in the biomass exceeds 80% [211]. Small yeast particles (preferably having a cross-section within the range of 0.2–2 mm) can be dried at the temperature of drying gas being up to 160 °C at the beginning of the process [214,215]. There is suggested a double-stage drying, the first stage being spray-drying, the second one — fluidized bed [211–213], or at the first stage using a conveyor drier, at the second one — a fluidized bed drier [216]. In the latter case, after the first stage of drying, which takes place up to moisture content 10–30%, the dried yeast is fractionated into small particles, less than 800 μm in size [216]. During drying the yeasts undergo the so-called "sticky stage", during which they have a marked tendency to agglomeration. This can be observed upon decrease from 50% to 20–30% moisture. Hence, by yeast drying in fluidized bed within the 20–50% range of moisture it is advisable to use disintegration forces, sufficient to prevent any substantial increase in particle size but insufficient to break the yeast cells [217–219]. The bulk of recent work advise to use various additives prior to drying — like hydrophobic silicic acid, and swelling, wetting and stabilizing agents [206,211–214,220,221]. Wetting agents most often are esters of sorbitol, glycerol or propylene glycol, fatty acids, and several other compounds, constituting 0.5–5% of yeast dry weight. Swelling agents can be methylcellulose, carboxymethyl cellulose and hydroxypropylcellulose (0.5–5%). In order to stabilize the emulsion it is advised to use arabic gum, guar gum, carrageenate (0.5–1%). The use of these compounds helps to obtain active dry yeast of a very high quality. Use of a protective additives like water emulsion of a mixture of food fat, esters of citric acid food fatty acid, mono- or diglycerol, ester of diacetyltartaric acid and food fatty acid monoglycerol, and mixtures there of has been proposed [222].

Since during rehydration of granulated dry yeast the restoration of cells is not simultaneous. The activity of the material may rather decrease. This is due to the fact that the exposed outer surface of the granules may rehydrate and swell and thus hinder the penetration of the moisture to the interior portions of the granule [223].

In order to cope with this drawback, there has been suggested a method for the production of porous granulated active dry yeasts. Moist yeasts to be dried are mixed with gas extruding the resultant mixture at a pressure of 1–10 atm through orifices and allowing escape of gas from the interior of the resultant extrudate to form porous granules of active yeast. Porous granules of dried yeasts can be rapidly rehydrated and restored, preserving high viability [223].

Cell damages and inactivation take place not only during dehydration, but also when storing dried preparations. Dried baker's yeast with 8–10% moisture content lost about 50% of their initial activity during 5 months [49]. Dried yeasts of 7.5–8.5% moisture, when stored in a semi-hermetic pack at room temperature, during the first months of storage lost about 7% of activity, and further on — slightly less [205]. It is considered that the lower the moisture content of preparations, the higher the viability of microbial cells during storage. For baker's yeast 7.0–8.5% is the optimum [15].

In order to increase the stability of dried yeasts during storage, it is advisable to add anti-oxidants, thiourea, sorbitane ether and some other additives into cells prior to their drying [56].

If an inert gas is introduced in the hermetic case, or vacuum is ensured, even 0.5% of oxygen may prove to be sufficient for diminishing yeast activity. In the USA dried yeast, meant for a long-term (1.0–1.5 years) storage at room temperature, is double packed — with polyethylene inside and aluminium foil on the outside. After packing and filling with nitrogen the cases are closed [224]. There have been suggested contemporary technologies for the activation of dried yeasts both, on media with various additives (wort concentrate, milk serum and ammonium salt) as well as by their heating at 30–60 °C [225,226].

The practice of dried microbial preparation storage has demonstrated that, notwithstanding the principal possibility to preserve the viability of dried cultures for decades, the practical term of storage for such preparations is only several years.

8 Conclusion

Avances in cytology, biochemistry and molecular biology in the last decades have widened our understanding of the dehydration-rehydration process. The review has discussed the main achievements in this field. To sum up the experimental results the main changes taking place in yeast cells during transition from active into anabiotic state and back are tressed.

Upon dehydration of yeasts to a 7–10% moisture content, almost all free and a part of bound water are removed. In yeasts when during drying a certain amount of isolated preserved water, is accumulated are most resistant to dehydration. The greatest damages of biopolymers and cell membranes take place at the final stage of dehydration, when a part of bound water is removed from the organisms. Various intracellular membranes undergo notable changes, there may locally appear a disorientation of bimolecular lipid layer in them. Dehydration increases membrane invaginations and may cause ruptures in its parts. The barrier function of the membrane is destroyed, by rehydration. This causes a leakage of electrolytes, carbohydrates, amino acids and nucleotides, as well as of macromolecules, and various enzymes. At the same time cell viability does nor correlate with the level of membrane permeability. Leakage of intracellular components from yeast cells may be significantly decreased during rehydration at increased (about 40 °C) temperatures, which is obviously connected with the change of phase transfer, temperature and intracellular membrane lipids.

Nucleic acids and proteins, too, undergo serious changes by dehydration. Desoxyribonucleic acids condense, a part of ribonucleic acids (up to 40%) may be subjected to degradation. Protein solubility decreases and they change their electrophoretic mobility. Enzyme activity may either increase or decrease during drying. Besides that, there is a possibility of *de novo* biosynthesis as well as a selective proteolysis. There has been shown an increase of fatty acid and saturation during dehydration. Yeasts with a high coefficient (3–4) of fatty acid desaturation were the most resistant to dehydration. It is interesting to note that reverse processes at work during rehydration. Trehalose content in cells, too, essentially affects yeast stability by dehydration. Yeasts with a 15–20% trehalose (d.m.) are the most suitable for drying. At present it seems unlikely that trehalose acts as a reserve carbohydrate, since its content, unlike that of glycogen, does not decrease during drying-rehydration. Polyphosphate content in yeasts, resistant to dehydration, is low, too. Hence polyphosphate and rehalose content in cells may serve as a criterion of their resistance. It is not in keeping with a former point of view, but active dry yeast can be obtained also from a nitrogen-rich biomass.

The present practice of baker's and wine yeast drying uses a variety of technologies and processes to obtain active dry yeast, which can preserve high commercial quality for a year and more. The present knowledge of yeast anabiosis, formed during the latest years on the basis of the information about changes, taking place in cells during their dehydration-rehydration, helps to establish the main reasons for the loss of viability in a part of dehydrated population.

Further studies in this field and a simultaneous elaboration of the equipment for commercial production of yeasts and their subsequent dehydration in order to obtain dry active preparations, as well as the development of new analytical equipment, may lead, and already in the nearest future, to the problem to how to make use of the accumulated knowledge in order to elaborate more advanced monitorable technological processes with the use of the most up-to-date sensors and electronic monitoring devices. A rapid determination of a number of "critical" parameters by respective monitoring computer equipment during the very process of cultivation may allow a swift change or optimization of the process in the direction, ensuring products of higher quality. The present experiments testify to the essential role, in this aspect, of the level of nucleic acid stability, trehalose and polyphosphate content, lipid fatty acid unsaturation degree, the state of intracellular membranes and the main cell organelles. At present all these data can still not be used in the monitoring of the processes, mainly because of the quite lengthy analytical procedures.

It will be possible however to forecast a significant improvement the quality of commercial dried baker's and wine yeasts in the next 5–10 years.

9 References

1. Preyer, W.: Die Erforschung des Lebens, Jena 1873
2. Schmidt, P.: Anabiosis, Moscow, USSR Acad. Sci. Publ. House 1955
3. Keilin, D.: Proc. Roy. Soc. London *B150*, 149 (1959)
4. Crowe, J. H., Clegg, J. S. (Eds.): Anhydrobiosis. Dowden, Hutchinson & Ross, Inc., Stroudsburg, Pennsylvania 1973
5. Pigón, A., Weglarska, B.: Bull. Acad. Pol. Sci. *1*, 69 (1953)
6. Pigón, A., Weglarska, B.: ibid. *3*, 31 (1955)
7. Plevako, J. A.: Dried Baker's Yeast, Pistchepromizdat, Moscow, 1953 (in Russian)
8. Crowe, J. H., Cooper, A. F.: Sci. Amer. *225*, 30 (1971)
9. Goldowski, A. M.: Anabiosis, Moscow, Nauka 1981 (in Russian)
10. Crowe, J. H., Clegg, J. S. (Eds.). Dry Biological Systems. Acad. Press, New York 1978
11. Womersley, C.: Comp. Biochem. Physiol. *70B*, 669 (1981)
12. Experimental Anabiosis, Riga, Zinatne 1985 (in Russian)
13. Makrushin, A. V.: Anhydrobiosis of Invertebrates, Leningrad, Nauka 1985 (in Russian)
14. Gould, G. W., Hurst, A. (Eds.): The Bacterial Spore, New York, Academic Press 1980[2]
15. Gerhardt, P., Costilow, R. N., Sadoff, H. L. (Eds.): Spores VI, American Society of Microbiology, Washington DC 1975
16. Chambliss, G., Vary, J. C. (Eds.): Spores VII, American Society of Microbiology, Washington DC 1978
17. Levinsen, H., et al. (Eds.): Sporulation and Germination, American Society of Microbiology, Washington DC 1981
18. Beker, M. J., Damberga, B. E., Rapoport, A. I.: Anabiosis of Microorganisms, Riga, Zinatne 1981 (in Russian)

19. Lozina-Lozinsky, L. K.: Zhurn. obschei biol. *34*, 153 (1973) (in Russian)
20. Meyer, G. H., Morrow, M. B., Wyss, O.: Nature *196*, 598 (1962)
21. Sneath, P. H. A.: ibid *195*, 643 (1962)
22. Hubalek, Z., Kockova-Kratochvilova, A.: Antonie van Leeuwenhoek, J. Microbiol. Serol. *44*, 229 (1978)
23. Hubalek, Z., Kockova-Kratochvilova, A.: Folia microbiol. *27*, 242 (1982)
24. Kockova-Kratochvilova, A., Hubalek, Z.: Antonie van Leeuwenhoek J. Microbiol. Serol. *49*, 571 (1983)
25. Wellman, A. M., Stewart, G. G.: Appl. Microbiol. *26*, 577 (1973)
26. Abyzov, S. S. et al.: Izvestiya AN SSSR, Ser. biol. 537 (1982) (in Russian)
27. Abyzov, S. S. et al.: ibid. 914 (1983) (in Russian)
28. Aksyonov, S. I.: Mikrobiologiya *51*, 877 (1982) (in Russian)
29. Gleeston, S., Leydler, K., Eiring, G.: Theory of Absolute Reaction Rates, Moscow, Izdat. inostr. lit. 1948 (in Russian)
30. Drozdova, T. B.: Amino Acid Geochemistry, Moscow, Nauka 1977 (in Russian)
31. Hare, P. E., Abelson, P. H.: Ann. rep. Director. Geophys. Lab. Carnegie Inst. 1966–1967, Year Book 66, 1968
32. Hare, P. E.: Geochemistry of proteins, peptides and amino acids, In: Organic Geochemistry. Methods and Results. B.—Heidelberg—New York, Springer-Verlag 1969
33. Drost-Hansen, W.: Structure and properties of water at biological interface, In: Chemistry of the Cell Interface, Pt. B (Ed. Brown, H. D.), p. 1. New York—London, Acad. Press 1971
34. Drost-Hansen, W.: Structure and functional aspects of interfacial (vicinal) water as related to membranes and cellular systems. Colloq. Internationaux du CNRS, N 246, Paris 1976
35. Etzler, F. M., Drost-Hansen, W.: A role for water in biological rate processes, In: Cell Associated Water (Eds. Drost-Hansen, W., Clegg, J. S.), p. 125. New York, Acad. Press 1979
36. Aksyonov, S. I.: Biofyzika *30*, 220 (1985) (in Russian)
37. Dick, D. A. T.: Cellwater, London, Butterworth 1966
38. Cope, J. H.: Bull. Math. Biophys. *29*, 583 (1967)
39. Schneider, M. J. T., Schneider, A. S.: J. Membr. Biol. *9*, 127 (1972)
40. Clegg, J. S.: Metabolism and the intracellular environment: the vicinal water network model, In: Cell Associated Water (Eds. Drost-Hansen, W., Clegg, J. S.), p. 363. New York, Acad. Press 1979
41. Gabyda, S. P.: Bound Water. Facts & Hypotheses, Novosibirsk, Nauka 1982 (in Russian)
42. Ostrovsky, D. N.: Molecular organization of biological membranes, In: Biomembranes. Structure, Functions, Investigation Methods, p. 7. Riga, Zinatne 1977 (in Russian)
43. Lykov, A. V.: Theory of Drying, Moscow—Leningrad, Gosenergoizdat 1950 (in Russian)
44. Beker, M. J.: Accelerated drying of yeasts in a suspended state and effect of main factors affecting the activity of dried yeasts, Diss., Riga 1959 (in Russian)
45. Beker, M. J., et al.: Appl. Microbiol. Biotechnol. *19*, 347 (1984)
46. Koga, S. H., Chigo, A., Nomura, K.: J. Biophys. *6*, 665 (1966)
47. Aksyonov, S. I., et al.: Studia biophysica *58*, 121 (1976)
48. Aksyonov, S. I., Goryachev, S. N., Nikolaev, G. M.: ibid. *85*, 15 (1981)
49. Beker, M. J.: Dehydration of Microbial Biomass, Riga, Zinatne 1967 (in Russian)
50. Rapoport, A. I., Medvedyeva, G. I.: Cytological investigations of yeast *Saccharomyces cerevisiae* resistance at freeze-drying, In: Microbial Preparations, p. 102, Riga, Zinatne 1976 (in Russian)
51. Harrison, J. S., Trevelyan, W. E.: Nature *200*, 1189 (1963)
52. Beker, M. J., Upitis, A. A.: Mikrobiologiya *41*, 830 (1972) (in Russian)
53. Popova, M. V.: Protein content and some dehydrogenase activity of yeast *Saccharomyces cerevisiae 14* depending on cultivation and dehydration conditions, Diss., Riga 1974 (in Russian)
54. Beker, M. J., et al.: Resistance and morphology at drying of yeast *Saccharomyces cerevisiae*, grown on molasses and ethanol media, In: Microbiological Preparations, p. 96. Riga, Zinatne 1976 (in Russian)
55. Beker, D., Grba, S.: Prehravib. Technol. Rev. *16*, 6 (1978)
56. Semikhatova, N. M.: Baker's Yeasts, Moscow, Pistchevaya promyshlennost 1980 (in Russian)

57. Rapoport, A. I., Beker, M. J.: Microbiology *52*, 556 (1983)
58. Litvinov, M. A.: Sov. Botanika *6–7*, 163 (1939) (in Russian)
59. Udeljnova, I. M.: Izv. AN SSSR, Ser. biol. *1*, 67 (1957) (in Russian)
60. Vas, K., Proszt, G.: Acta Microbiol. Acad. Sci. Hung. *6*, 283 (1959)
61. Rapoport, A. I., Biryuzova, V. I., Meissel, M. N.: Dokladi AN SSSR *213*, 708 (1973) (in Russian)
62. Biryuzova, V. I., et al.: Membrane structure of yeast cells, their origin and rearrangement under various physiological conditions, In: Proc. Fourth Intern. Symp. on Yeasts, p. 191, Vienna 1974
63. Rapoport, A. I., et al.: Ultrastructural organization of yeast cells in anabiotic state, In: Proc. XV Czechosl. Conf. on Electron Microscopy with Intern. Partic., P. A 257, Prague 1977
64. Rapoport, A. I.: Ultrastructural changes in yeast cells at their transition into anabiotic state, In: Biotechnology and Bioengineering, II Media and Products of Microbial Synthesis, p. 138. Riga, Zinatne 1978 (in Russian)
65. Rapoport, A. I., et al.: Ultrastructure studies of yeast organism anabiosis, In: Electron Microscopy in Botanical Investigations, p. 221. Riga, Zinatne 1978 (in Russian)
66. Galeotti, C. L., Williams, K. L.: J. Gen. Microbiol. *104*, 337 (1978)
67. Rapoport, A. I.: Some results of a luminiscent microscopy studies of dried yeast cells, In: Microbiological Synthesis of Amino Acids, p. 57. Riga, Zinatne 1977 (in Russian)
68. Rapoport, A. I., Kostrikina, N. A.: Izvestiya AN SSSR, Ser. biol. *5*, 770 (1973) (in Russian)
69. Rapoport, A. I., Ventina, E. J.: Basic cytological regularities, taking place at yeast organism transfer into anabiosis, In: XIth All-Union Conference on Electron Microscopy, II Biology, p. 73, Moscow, Nauka 1979 (in Russian)
70. Beker, M. J., et al.: Biochemical and morphological peculiarities of yeast resistant towards dehydration, In: Abstr. Fifth Intern. Ferment. Symp. p. 491, Berlin 1976
71. Biryuzova, V. I.: Membrane Structures of Microorganisms, Moscow, Nauka 1973 (in Russian)
72. Petrikevich, S. B.: Accumulation, distribution and transformations of aromatic carbohydrate 3,4-benzpyrene in microbial cells, Diss., Moscow 1965 (in Russian)
73. Biryuzova, V. I., Rapoport, A. I.: Mikrobiologiya *47*, 300 (1978) (in Russian)
74. Beker, M. J., et al.: Factors of wine yeast resistance at their dehydration, In: Current Development in Yeast Research. Advances in Biotechnology (Eds. Stewart, G. G., Russel, J.), p. 117. Toronto—Oxford—New York—Sydney—Paris—Frankfurt, Pergamon Press 1981
75. Ventiņa, E. J., et al.: Mikrobiologiya *53*, 658 (1984) (in Russian)
76. Rapoport, A. I., et al.: Use of optical-structural computer analysis in the study of yeast organism anabiosis, In: Automatic Systems for Image Treatment, p. 261. Moscow, Nauka 1981 (in Russian)
77. Rapoport, A. I., Beker, M. J.: Mikrobiologiya *54*, 450 (1985) (in Russian)
78. Rapoport, A. I., Biryuzova, V. I., Beker, M. J.: Microbiology *52*, 204 (1983)
79. Blumbergs, J. E., et al.: Damage and reparation of yeast cytoplasmic membrane at dehydration and subsequent reactivation, In: Vth Intern. Biophysics Congress of the Intern. Union for Pure and Appl. Biophysics, p. 90. Copenhagen 1975
80. Ventina, E. J.: Cytoplasmic membrane during the process of yeast *Saccharomyces cerevisiae* dehydration, In: Biotechnology and Bioengineering, II Media and Products of Microbial Synthesis, p. 37. Riga, Zinatne 1978 (in Russian)
81. Ventiņa, E. J., Salulite, L. A., Rapoport, A. I.: Some ultrastructure peculiarities of yeast cell development and their reaction to dehydration and subsequent reactivation processes, In: Plant Ultrastructure, Kishinev 1983 (in Russian)
82. Ventina, E. J., Saulite, L. A., Rapoport, A. I.: Effect of dried cell rehydration on their ultrastructure and viability, In: Microbial Synthesis of Enzymes and Obtaining of Their Preparative Forms, p. 59. Riga, Zinatne 1983 (in Russian)
83. Nasonov, D. N.: Local Reaction of Protoplasm and Spreading Agitation, Moscow—Leningrad, Izd-vo AN SSSR 1959 (in Russian)
84. Rapoport, A. I., Markovsky, A. B., Beker, M. J.: Microbiology *51*, 707 (1982)
85. Meissel, M. N., et al.: Izv. AN SSSR, Ser. biol. *6*, 827 (1964) (in Russian)
86. Baird-Parker, A. C., Davenport, E.: J. Appl. Bacteriol. *28*, 390 (1965)
87. Tomlins, R. J., Ordal, Z. J.: J. Bacteriol. *105*, 512 (1971)

88. Sinskey, T. J., Silverman, G. J.: ibid. *101*, 429 (1970)
89. Katsui, N., et al.: ibid. *151*, 1523 (1982)
90. Rapoport, A. I.: Mikrobiologiya *42*, 357 (1973) (in Russian)
91. Biryuzova, V. J., Volkova, T. M.: Intrazellulare Membranstrukturen der Hefeorganismen, In: Symp. on Electron Microscopy II, p. 299. Tokyo 1966
92. Ventiņa, E. J., Rapoport, A. I.: Cytological reactions of yeast organisms at transition into anabiosis and subsequent rehydration, In: Experimental Studies on Microorganism Development, p. 115, Puschino-on-the-Oka 1978 (in Russian)
93. Novichkova, A. T., Rapoport, A. I.: Microbiology *53*, 5 (1984)
94. Herrera, T., et al.: Arch. Biochem. Biophys. *63*, 131 (1956)
95. Suomalainen, H., Oura, E., Linnahalme, T.: J. Inst. Brew. *71*, 330 (1965)
96. Ramnietse, V. E., Skárds, I. V., Popova, M. V.: Mikrobiologiya *47*, 430 (1978) (in Russian)
97. Altman, K. I., Gerber, G., Okada, S.: Radiation Biochemistry, vol. 1, Cells, New York—London, Acad. Press 1970
98. Popova, M. V., et al.: Microbiology *50*, 671 (1981)
99. Beker, M. J.: Izv. AN LatvSSR *5*, 23 (1978) (in Russian)
100. Auzan, S. I., et al.: Permeability of dried yeast *Saccharomyces cerevisiae 14*, In: Biotechnology and Bioengineering, II Media and Products of Microbial Synthesis, p. 15. Riga, Zinatne 1978 (in Russian)
101. Mowshowitz, D. B.: Anal. Biochem. *70*, 94 (1976)
102. Singer, S. J., Conrad, M. J.: The structure of cell membranes, In: Cell Membrane Funct. and Disfunct. Vascular Tissue, Proc. 5th Argent. Symp., Waterloo, 30 June–1 July, 1980, p. 1. Amsterdam e. a. 1981
103. Shimshick, E., McConell, H. M.: Biochemistry *12*, 235 (1973)
104. Schindler, M., Osborn, M., Koppel, D.: Nature *283*, 346 (1980)
105. Konev, S. V., Mazhulj, V. M.: Intercellular Contacts, Minsk, Nauka i Tekhnika 1977
106. Nurminen, T., Raskinen, L., Suomalainen, H.: J. Gen. Microbiol. *98*, 301 (1977)
107. Kaprelyants, A. S.: Biochemistry *47*, 739 (1982)
108. Jost, P., Nadakavukaren, K., Griffith, O.: ibid. *16*, 3111 (1977)
109. Engelhardt, V., Glaser, M., Storm, D.: ibid *17*, 3191 (1978)
110. Marcelja, S.: Biochim. Biophys. Acta *455*, 1 (1976)
111. Vanderkooi, C., Bendler, J. T.: Dynamics and thermodynamics of lipid-protein interactions in membranes, In: Structure of Biological Membranes (Eds. Abrahansons, S., Pascher, J.), p. 551. New York—London, Plenum Press 1977
112. Dergunov, A. D., et al.: Biochemistry *47*, 250 (1982)
113. Farias, R. N., et al.: Biochim. Biophys. Acta *415*, 231 (1975)
114. Crowe, J. H., O'Dell, S. J., Armstrong, D. A.: J. Exp. Zool. *297*, 431 (1979)
115. Luzzati, V., Husson, F.: J. Cell Biol. *12*, 207 (1962)
116. Finean, J. B., Coleman, E., Michel, R. H.: Membranes and their Cellular Functions, Oxford—London—Edinburgh—Melbourne, Blackwell Sci. Publ. 1974
117. Shipley, G. G.: Recent X-ray diffraction studies of biological membranes, In: Biological Membranes, vol. 2, p. 1. London—New York 1973
118. Beker, M. J.: Microorganism biomembranes at dehydration, rehydration and reactivation, In: Biomembranes, Structure, Functions, Investigation Methods, p. 216. Riga, Zinatne 1977 (in Russian)
119. Beker, M. J., Rapoport, A. I., Ventiņa, E. J.: Convective drying as a damaging factor of cell structures, In: General Mechanisms of Cellular Reactions to Damaging Effects, p. 21. Leningrad 1977 (in Russian)
120. Simon, E. W.: Membranes in dry and imbibing seeds, In: Dry Biological Systems (Eds. Crowe, J. H., Clegg, J. S.), p. 205. New York, Acad. Press 1978
121. Crowe, J. H., Crowe, L. M.: Cryobiology *19*, 317 (1982)
122. Gordon-Kamm, W. J., Steponkus, P. L.: Proc. Natl. Acad. Sci. USA *81*, 6373 (1984)
123. Singh, J., et al.: Plant Physiol. *75*, 1075 (1984)
124. Gordon, D. E., Curnutte, B., Lark, K. G.: J. Mol. Biol. *13*, 571 (1965)
125. Sutherland, G. B. B. M., Tsuboi, N.: Proc. Roy. Soc. A. *239*, 446 (1957)
126. Bradbury, E. M., Price, W. C., Wilkinson, G. R.: J. Mol. Biol. *3*, 301 (1961)

127. Spitkovsky, D. M., Tseitlin, P. I., Tongur, V. S.: Biofizika *5*, 3 (1960) (in Russian)
128. Ando, Y., Fukada, E.: J. Radiat. Res. *15*, 212 (1974)
129. Servin-Massieu, M.: Curr. Top. Microbiol. Immunol. *54*, 119 (1971)
130. Hieda, K.: Mutat. Res. *84*, 17 (1981)
131. Hieda, K.: J. Bacteriol. *150*, 963 (1982)
132. Rapoport, A. I., et al.: Microbiology *50*, 163 (1981)
133. Rapoport, A. I., et al.: Anabiotic state of yeast organisms, In: Yeasts in Human Environment. IX Intern. Spec. Symp. on Yeasts. Abstracts, p. 44. Smolenice 1983
134. Bachmann, B., Kosiek, E.: Mitteil. Versuchsstat. Gärungsgewerbe *27*, 45 (1973)
135. Damberga, B. E., Upitis, A. A.: Effect of dehydration and storage on amino nitrogen content in carbohydrates in yeasts, In: Anabiosis and Preanabiosis of Microorganisms, p. 63. Riga, Zinatne 1973 (in Russian)
136. Beker, M. J., et al.: Dehydration as a regulating factor of yeast metabolism, In: FEMS International Symposium. Environmental Regulation of Microbial Metabolism. Abstracts, p. 20. Puschino 1983
137. Popova, M. V., Damberga, B. E.: Izv. AN Latv. SSR *9*, 131 (1972) (in Russian)
138. Popova, M. V., Damberga, B. E.: ibid. *11*, 75 (1973) (in Russian)
139. Popova, M. V., Damberga, B. E.: Mikrobiologiya *44*, 1005 (1975) (in Russian)
140. Kulaev, S. I., et al.: Prikl. Biokhim. Mikrobiol. *13*, 893 (1977) (in Russian)
141. Rapoport, A. I., Beker, M. J.: Microbiology *47*, 136 (1978)
142. Laivenieks, M. G., et al.: Prikl. Biokhim. Mikrobiol. *14*, 690 (1978) (in Russian)
143. Laivenieks, M. G., et al.: Eur. J. Appl. Microbiol. Biotechnol. *10*, 349 (1981)
144. Brivkalne, I. K., Damberga, B. E.: Biochemical properties and respiratory activity of yeast *Saccharomyces cerevisiae 14* at changed cultivation and rehydration conditions, In: Microbiological Preparations, p. 86. Riga, Zinatne 1976 (in Russian)
145. Auzan, S. I., Véze, V. A.: Thermostable properties of invertase in *Saccharomyces cerevisiae 14*, In: Physiology of Epiphytic and Root Microorganisms, p. 134. Riga, Zinatne 1979 (in Russian)
146. Proskuryakov, N. I., Oparicheva, E. F.: Mikrobiologiya *25*, 600 (1956) (in Russian)
147. Fulco, A. I.: Ann. Rev. Biochem. *43*, 215 (1974)
148. Sinensky, M.: Proc. Natl. Acad. Sci. USA *71*, 522 (1974)
149. Sandermann, H.: Biochim. Biophys. Acta *515*, 209 (1978)
150. Hall, M. J., Ratledge, C.: Appl. Environ. Microbiol. *33*, 577 (1977)
151. Ratledge, C., Botham, P. A.: J. Gen. Microbiol. *102*, 391 (1977)
152. Botham, P. A., Ratledge, C.: ibid. *114*, 361 (1979)
153. Boulton, C. A., Ratledge, C.: ibid. *127*, 169 (1981)
154. Boulton, C. A., Ratledge, C.: ibid. *127*, 423 (1981)
155. Boulton, C. A., Ratledge, C.: ibid. *129*, 2863 (1983)
156. Boulton, C. A., Ratledge, C.: ibid. *129*, 2871 (1983)
157. Keenan, M. H. J., Rose, A. H.: FEMS Microbiol. Lett. *6*, 133 (1979)
158. Rose, A. H.: Osmotic stress and microbial survival, In: Surv. Vegetative Microbes. 26th Symp. Soc. Gen. Microbiol., Univ. Cambridge, 1976, p. 155. Cambridge e. a. 1976
159. Hossack, J. A., Belk, D. M., Rose, A. H.: Arch. Microbiol. *114*, 137 (1977)
160. Hossack, J. A., Sharpe, V. J., Rose, A. H.: J. Bacteriol. *129*, 1144 (1977)
161. Pringle, A. T., Rose, A. H.: J. Gen. Microbiol. *111*, 227 (1979)
162. Walton, E. T., Pringle, J. R., Arch. Microbiol. *124*, 285 (1980)
163. Calcott, P. H., Rose, A. H.: J. Gen. Microbiol. *128*, 549 (1982)
164. Hossack, J. A., Rose, A. H., Dawson, P. S. S.: ibid. *113*, 199 (1979)
165. Watson, K., Rose, A. H.: ibid. *117*, 225 (1980)
166. Beker, M. J., et al.: Biomembranes of yeast during dehydration-rehydration, In: Abstr. VIth Intern. Spec. Symp. on Yeast, p. 2. Montpellier 1978
167. Auziņa, L. P., et al.: Prikl. Biokhim. i Mikrobiol. *15*, 822 (1979) (in Russian)
168. Zikmanis, P. B., et al.: Eur. J. Appl. Microbiol. Biotechnol. *15*, 100 (1982)
169. Zikmanis, P. B., et al.: Izv. AN LatvSSR *1*, 80 (1983) (in Russian)
170. Zikmanis, P. B., et al.: The studies of yeast *Saccharomyces cerevisiae* fatty acid content and glycolytic pathway changes at drying-rehydration, In: Yeast Technology: in Focus for the Future, VIIIth Intern. Symp. on Yeasts, Abstracts of Papers Accepted, p. A125. Bombay 1983

171. Zikmanis, P. B., et al.: Eur. J. Appl. Microbiol. Biotechnol. *18*, 298 (1983)
172. Golubeva, N. P., Dubinskaya, A. V., Timofeyeva, S. V.: Mikrobiol. promyshlennostj *4*, 51 (1970) (in Russian)
173. Saubenova, M. G.: Polysaccharides of Yeast Organisms, Alma-Ata 1976 (in Russian)
174. Beker, M. J., et al.: Structure and function of the yeast *Saccharomyces cerevisiae* cellular envelope at dehydration-rehydration, In: VIIth Intern. Spec. Symp. on Yeast: Abstracts, p. 80, Valencia 1981
175. Notkina, L. G.: Trudi Instituta Mikrobiologii AN SSSR *6*, 183 (1959) (in Russian)
176. Clegg, J. S.: Comp. Biochem. Physiol. *20*, 8–1 (1967)
177. Crowe, J. H., Madin, K. A. C.: J. Exp. Zool. *193*, 323 (1975)
178. Paven, R.: Canad. J. Res. *27*, 749 (1949)
179. Loomis, S. H., O'Dell, S. J., Crowe, J. H.: J. Exp. Zool. *208*, 355 (1979)
180. Crowe, J. H., Crowe, L. M., Chapman, D.: Science *223*, 701 (1984)
181. Roth, R.: J. Bacteriol. *101*, 53 (1970)
182. Damberga, B. E.: Izv. AN LatvSSR *8*, 77 (1964) (in Russian)
183. Damberga, B. E.: ibid *8*, 93 (1982) (in Russian)
184. Thevelein, J. M., den Hollander, J. A., Schulman, R. G.: Arch. Int. Physiol. Biochim. *90*, B70 (1982)
185. Beker, M. J., et al.: Microbiology *43*, 874 (1974)
186. Kulaev, I. S.: The Biochemistry of Inorganic Polyphosphates, A Whiley-Interscience Publication, Chichester—New York—Brisbane—Toronto 1979
187. Ventiņa, E. J., et al.: Zbl. Mikrobiol. *139*, 307 (1984)
188. Zikmanis, P. B., et al.: Mikrobiologiya *54*, 406 (1985) (in Russian)
189. Leach, R. H., Scott, W. J.: J. Gen. Microbiol. *21*, 195 (1959)
190. Herrera, T., et al.: Arch. Biochem. Biophys. *63*, 131 (1956)
191. Leder, I. G.: J. Bacteriol. *111*, 211 (1972)
192. Levin, S. V.: Structural changes in membrane structures. Leningrad, Nauka 1976
193. Overäht, P., Träuble, H.: Biochemistry *12*, 2625 (1973)
194. McDonald, R. C., Simon, S. A., Baer, E.: ibid. *15*, 885 (1976)
195. Veksli, Z., Salsbury, N. J., Chapman, D.: Biochem. Biophys. Acta *193*, 434 (1969)
196. Record, B. R., Taylor, R., Miller, D. S.: J. Gen. Microbiol. *28*, 585 (1962)
197. Ray, B., Jezeski, J. J., Busta, F. F.: Appl. Microbiol. *22*, 184 (1971)
198. Ray, B., Jezeski, J. J., Busta, F. F.: ibid. *22*, 401 (1971)
199. Berestovsky, G. N.: Biofizika *20*, 633 (1975) (in Russian)
200. Levine, J. K.: Progress in Biophys. and Molec. Biol. *24*, 1 (1972)
201. Hauser, H., Philip, M. C., Berrat, M. D.: Biochem. Biophys. Acta *413*, 341 (1975)
202. Naftalin, R. J., Symons, M. C. R.: ibid. *352*, 173 (1974)
203. Korogodin, V. I.: Problems of Postradiation Restoration, Moscow, Atomizdat 1966 (in Russian)
204. Fateyeva, M. V.: Methods of yeast culture collection storage, In: Methods of Microorganism Culture Collection Storage, Moscow, Nauka 1967 (in Russian)
205. Reed, G., Peppler, H.: Yeast Technology, Westport, AVI Publ. Co. 1973
206. USA Pat. N 4370420 (1983)
207. Jap. Pat. N 586470 (1983)
208. FRG Pat. N 3225970 (1983)
209. FRG Pat. N 2515029 (1978)
210. FRG Pat. N 2559462 (1979)
211. Great Britain Pat. N 1459085 (1976)
212. Great Britain Pat. N 1459211 (1976)
213. Great Britain Pat. N 1459407 (1976)
214. USA Pat. N 4217420 (1980)
215. USA Pat. N 4341871 (1982)
216. FRG Pat. N 2147715 (1979)
217. Great Britain Pat. N 1498301 (1978)
218. Great Britain Pat. N 1498302 (1978)
219. USA Pat. N 4188407 (1980)
220. Great Britain Pat. N 1321714 (1973)

221. Great Britain Pat. N 1539211 (1979)
222. FRG Pat. N 3320654 (1984)
223. USA Pat. N 4335144 (1982)
224. Dale R. F.: Mod. Packing *29*, 170 (1956)
225. USSR Pat. N 1051119 (1983)
226. USA Pat. N 4397877 (1983)

Author Index Volumes 1–35

Acosta Jr., D. see Smith, R. V. Vol. 5, p. 69

Acton, R. T., Lynn, J. D.: Description and Operation of a Large-Scale Mammalian Cell, Suspension Culture Facility. Vol. 7, p. 85

Agrawal, P., Lim, H. C.: Analysis of Various Control Schemes for Continuous Bioreactors. Vol. 30, p. 61

Aiba, S.: Growth Kinetics of Photosynthetics Microorganisms. Vol. 23, p. 85

Aiba, S., Nagatani, M.: Separation of Cells from Culture Media. Vol. 1, p. 31

Aiba, S. see Sudo, R. Vol. 29, p. 117

Aiba, S., Okabe, M.: A Complementary Approach to Scale-Up. Vol. 7, p. 111

Alfermann, A. W. see Reinhard, E. Vol. 16, p. 49

Anderson, L. A., Phillipson, J. D., Roberts, M. F.: Biosynthesis of Secondary Products by Cell Cultures of Higher Plants. Vol. 31, p. 1

Arnaud, A. see Jallageas, J.-C. Vol. 14, p. 1

Arora, H. L., see Carioca, J. O. B. Vol. 20, p. 153

Asher, Z. see Kosaric, N. Vol. 32, p. 25

Atkinson, B., Daoud, I. S.: Microbial Flocs and Flocculation. Vol. 4, p. 41

Atkinson, B., Fowler, H. W.: The Significance of Microbial Film in Fermenters. Vol. 3, p. 221

Bagnarelli, P., Clementi, M.: Serum-Free Growth of Human Hepatoma Cells. Vol. 34, p. 85

Barker, A. A., Somers, P. J.: Biotechnology of Immobilized Multienzyme Systems. Vol. 10, p. 27

Beardmore, D. H. see Fan, L. T. Vol. 14, p. 101

Bedetti, C., Cantafora, A.: Extraction and Purification of Arachidonic Acid Metabolites from Cell Cultures. Vol. 35, p. 47

Belfort, G. see Heath, C. Vol. 34, p. 1

Beker, M. J., Rapoport, A. J.: Conservation of Yeasts by Dehydration. Vol. 35, p. 127

Bell, D. J., Hoare, M., Dunnill, P.: The Formation of Protein Precipitates and their Centrifugal Recovery. Vol. 26, p. 1

Berlin, J., Sasse, F.: Selection and Screening Techniques for Plant Cell Cultures. Vol. 31, p. 99

Binder, H. see Wiesmann, U. Vol. 24, p. 119

Bjare, M.: Serum-Free Cultivation of Lymphoid Cells. Vol. 34, p. 95

Blanch, H. W., Dunn, I. J.: Modelling and Simulation in Biochemical Engineering. Vol. 3, p. 127

Blanch, H. W., see Moo-Young, M. Vol. 19, p. 1

Blanch, H. W., see Maiorella, B. Vol. 20, p. 43

Blenke, H. see Seipenbusch, R. Vol. 15, p. 1

Blenke, H.: Loop Reactors. Vol. 13, p. 121

Blumauerová, M. see Hostalek, Z. Vol. 3, p. 13

Böhme, P. see Kopperschläger, G. Vol. 25, p. 101

Bottino, P. J. see Gamborg, O. L. Vol. 19, p. 239
Bowers, L. D., Carr, P. W.: Immobilized Enzymes in Analytical Chemistry. Vol. 15, p. 89
Brauer, H.: Power Consumption in Aerated Stirred Tank Reactor Systems. Vol. 13, p. 87
Brodelius, P.: Industrial Applications of Immobilized Biocatalysts. Vol. 10, p. 75
Brosseau, J. D. see Zajic, J. E. Vol. 9, p. 57
Bryant, J.: The Characterization of Mixing in Fermenters. Vol. 5, p. 101
Buchholz, K.: Reaction Engineering Parameters for Immobilized Biocatalysts. Vol. 24, p. 39
Bungay, H. R.: Biochemical Engineering for Fuel Production in United States. Vol. 20, p. 1
Butler, M.: Growth Limitations in Microcarriers Cultures. Vol. 34, p. 57

Cantafora, A. see Bedetti, C. Vol. 35, p. 47
Chan, Y. K. see Schneider, H. Vol. 27, p. 57
Carioca, J. O. B., Arora, H. L., Khan, A. S.: Biomass Conversion Program in Brazil. Vol. 20, p. 153
Carr, P. W. see Bowers, L. D. Vol. 15, p. 89
Chang, M. M., Chou, T. Y. C., Tsao, G. T.: Structure, Preteatment, and Hydrolysis of Cellulose. Vol. 20, p. 15
Charles, M.: Technical Aspects of the Rheological Properties of Microbial Cultures. Vol. 8, p. 1
Chen, L. F., see Gong, Ch.-S. Vol. 20, p. 93
Chou, T. Y. C., see Chang, M. M. Vol. 20, p. 15
Cibo-Geigy/Lepetit: Seminar on Topics of Fermentation Microbiology. Vol. 3, p. 1
Claus, R. see Haferburg, D. Vol. 33, p. 53
Clementi, P. see Bagnarelli, P. Vol. 34, p. 85
Cogoli, A., Tschopp, A.: Biotechnology in Space Laboratories. Vol. 22, p. 1
Cooney, C. L. see Koplove, H. M. Vol. 12, p. 1
Costentino, G. P. see Kosaric, N. Vol. 32, p. 1

Daoud, I. S. see Atkinson, B. Vol. 4, p. 41
Das, K. see Ghose, T. K. Vol. 1, p. 55
Davis, P. J. see Smith, R. V. Vol. 14, p. 61
Deckwer, W.-D. see Schumpe, A. Vol. 24, p. 1
Demain, A. L.: Overproduction of Microbial Metabolites and Enzymes due to Alteration of Regulation. Vol. 1, p. 113
Doelle, H. W., Ewings, K. N., Hollywood, N. W.: Regulation of Glucose Metabolism in Bacterial Systems. Vol. 23, p. 1
Dunn, I. J. see Blanch, H. W. Vol. 3, p. 127
Dunnill, P. see Bell, D. J. Vol. 26, p. 1
Duvnjak, Z., see Kosaric, N. Vol. 20, p. 119
Duvnjak, Z. see Kosaric, N. Vol. 32, p. 1

Eckenfelder Jr., W. W., Goodman, B. L., Englande, A. J.: Scale-Up of Biological Wastewater Treatment Reactors. Vol. 2, p. 145
Einsele, A., Fiechter, A.: Liquid and Solid Hydrocarbons. Vol. 1, p. 169
Electricwala, A. see Griffiths, J. B. Vol. 34, p. 147
Enari, T. M., Markkanen, P.: Production of Cellulolytic Enzymes by Fungi. Vol. 5, p. 1
Enatsu, T., Shinmyo, A.: In Vitro Synthesis of Enzymes. Physiological Aspects of Microbial Enzyme Production Vol. 9, p. 111
Englande, A. J. see Eckenfelder Jr., W. W. Vol. 2, p. 145

Eriksson, K. E.: Swedish Developments in Biotechnology Based on Lignocellulose Materials. Vol. 20. p. 193

Esser, K.: Some Aspects of Basic Genetic Research on Fungi and Their Practical Implications. Vol. 3, p. 69

Esser, K., Lang-Hinrichs, Ch.: Molecular Cloning in Heterologous Systems, Vol. 26, p. 143

Ewings, K. N. see Doelle, H. W. Vol. 23, p. 1

Faith, W. T., Neubeck, C. E., Reese, E. T.: Production and Application of Enzymes. Vol. 1, p. 77

Fan, L. S. see Lee, Y. H. Vol. 17, p. 131

Fan, L. T., Lee, Y.-H., Beardmore, D. H.: Major Chemical and Physical Features of Cellulosic Materials as Substrates for Enzymatic Hydrolysis. Vol. 14, p. 101

Fan, L. T., Lee, Y.-H., Gharpuray, M. M.: The Nature of Lignocellulosics and Their Pretreatments for Enzymatic Hydrolysis. Vol. 23, p. 155

Fan, L. T. see Lee, Y.-H. Vol. 17, p. 101 and p. 131

Faust, U., Sittig, W.: Methanol as Carbon Source for Biomass Production in a Loop Reactor. Vol. 17, p. 63

Fiechter, A.: Physical and Chemical Parameters of Microbial Growth. Vol. 30, p. 7

Fiechter, A. see Einsele, A. Vol. 1, p. 169

Fiechter, A. see Janshekar, H. Vol. 27, p. 119

Finocchiaro, T., Olson, N. F., Richardson, T.: Use of Immobilized Lactase in Milk Systems. Vol. 15, p. 71

Flaschel, E. see Wandrey, C. Vol. 12, p. 147

Flaschel, E., Wandrey, Ch., Kula, M.-R.: Ultrafiltration for the Separation of Biocatalysts. Vol. 26, p. 73

Flickinger, M. C., see Gong, Ch.-S. Vol. 20, p. 93

Fowler, H. W. see Atkinson, B. Vol. 3, p. 221

Fukui, S., *Tanaka, A.*: Application of Biocatalysts Immobilized by Prepolymer Methods. Vol. 29, p. 1

Fukui, S., *Tanaka, A.*: Metabolism of Alkanes by Yeasts. Vol. 19, p. 217

Fukui, S., Tanaka, A.: Production of Useful Compounds from Alkane Media in Japan, Vol. 17, p. 1

Galzy, P. see Jallageas, J.-C. Vol. 14, p. 1

Gamborg, O. L., Bottino, P. J.: Protoplasts in Genetic Modifications of Plants. Vol. 19, p. 239

Gaudy Jr., A. F., Gaudy, E. T.: Mixed Microbial Populations. Vol. 2, p. 97

Gaudy, E. T. see Gaudy Jr., A. F. Vol. 2, p. 97

Gharpuray, M. M. see Fan, L. T. Vol. 23, p. 155

Ghose, T. K., Das, K.: A Simplified Kinetic Approach to Cellulose-Cellulase System. Vol. 1, p. 55

Ghose, T. K.: Cellulase Biosynthesis and Hydrolysis of Cellulosic Substances. Vol. 6, p. 39

Gogotov, I. N. see Kondratieva, E. N. Vol. 28, p. 139

Gomez, R. F.: Nucleic Acid Damage in Thermal Inactivation of Vegetative Microorganisms. Vol. 5, p. 49

Gong, Ch.-S. see McCracken, L. D. Vol. 27, p. 33

Gong, Ch.-S., Chen, L. F., Tsao, G. T., Flickinger, M. G.: Conversion of Hemicellulose Carbohydrates, Vol. 20, p. 93

Goodman, B. L. see Eckenfelder Jr., W. W. Vol. 2, p. 145

Graves, D. J., Wu, Y.-T.: The Rational Design of Affinity Chromatography Separation Processes. Vol. 12, p. 219

Griffiths, J. B., Electricwala, A.: Production of Tissue Plasminogen Activators from Animal Cells. Vol. 34, p. 147

Gutschick, V. P.: Energetics of Microbial Fixation of Dinitrogen. Vol. 21, p. 109

Haferberg, D., Hommel, R., Claus, R., Kleber, H.-P.: Extracellular Microbial Lipids as Biosurfactants. Vol. 33, p. 53

Hahlbrock, K., Schröder, J., Vieregge, J.: Enzyme Regulation in Parsley and Soybean Cell Cultures, Vol. 18, p. 39

Haltmeier, Th.: Biomass Utilization in Switzerland. Vol. 20, p. 189

Hampel, W.: Application of Microcomputers in the Study of Microbial Processes. Vol. 13, p. 1

Harder, A., Roels, J. A.: Application of Simple Structured Models in Bioengineering. Vol. 21, p. 55

Harrison, D. E. F., Topiwala, H. H.: Transient and Oscillatory States of Continuous Culture. Vol. 3, p. 167

Heath, C., Belfort, C.: Immobilization of Suspended Mammalian Cells: Analysis of Hollow Fiber and Microcapsule Bioreactors. Vol. 34, p. 1

Hedman, P. see Janson, J.-C. Vol. 25, p. 43

Heinzle, E.: Mass Spectrometry for On-line Monitoring of Biotechnological Processes. Vol. 35, p. 1

Ho, Ch., Smith, M. D., Shanahan, J. F.: Carbon Dioxide Transfer in Biochemical Reactors. Vol. 35, p. 83

Hoare, M. see Bell, D. J. Vol. 26, p. 1

Hofmann, E. see Kopperschläger, G. Vol. 25, p. 101

Holló, J. see Nyeste, L. Vol. 26, p. 175

Hollywood, N. W. see Doelle, H. W. Vol. 23, p. 1

Hommel, R. see Haferburg, D. Vol. 33, p. 53

Hošiálek, Z., Blumauerová, M., Vanek, Z.: Genetic Problems of the Biosynthesis of Tetracycline Antibiotics. Vol. 3, p. 13

Hu, G. Y. see Wang, P. J. Vol. 18, p. 61

Humphrey, A. E., see Rolz, G. E. Vol. 21, p. 1

Hustedt, H. see Kula, M.-R. Vol. 24, p. 73

Imanaka, T.: Application of Recombinant DNA Technology to the Production of Useful Biomaterials. Vol. 33, p. 1

Inculet, I. I. see Zajic, J. E. Vol. 22, p. 51

Jack, T. R., Zajic, J. E.: The Immobilization of Whole Cells. Vol. 5, p. 125

Jallageas, J.-C., Arnaud, A., Galzy, P.: Bioconversions of Nitriles and Their Applications. Vol. 14, p. 1

Jang, C.-M., Tsao, G. T.: Packed-Bed Adsorption Theories and Their Applications to Affinity Chromatography. Vol. 25, p. 1

Jang, C.-M., Tsao, G. T.: Affinity Chromatography. Vol. 25, p. 19

Jansen, N. B., Tsao, G. T.: Bioconversion of Pentoses to 2,3-Butanediol by Klebsiella pneumonia. Vol. 27, p. 85

Janshekar, H., Fiechter, A.: Lignin Biosynthesis, Application, and Biodegradation. Vol. 27, p. 119

Janson, J.-C., Hedman, P.: Large-Scale Chromatography of Proteins. Vol. 25, p. 43

Jeffries, Th. W.: Utilization of Xylose by Bacteria, Yeasts, and Fungi. Vol. 27, p. 1

Jiu, J.: Microbial Reactions in Prostaglandin Chemistry, Vol. 17, p. 37

Kamihara, T., Nakamura, I.: Regulation of Respiration and Its Related Metabolism by Vitamin B_1 and Vitamin B_6 in Saccharomyces Yeasts. Vol. 29, p. 35

Keenan, J. D. see Shieh, W. K. Vol. 33, p. 131

Khan, A. S., see Carioca, J. O. B. Vol. 20, p. 153

Kimura, A.: Application of recDNA Techniques to the Production of ATP and Glutathione by the "Syntechno System". Vol. 33, p. 29

King, C.-K. see Wang, S. S. Vol. 12, p. 119

King, P. J.: Plant Tissue Culture and the Cell Cycle, Vol. 18, p. 1

Kjaergaard, L.: The Redox Potential: Its Use and Control in Biotechnology. Vol. 7, p. 131

Kleber, H.-P. see Haferburg, D. Vol. 33, p. 53

Kleinstreuer, C., Poweigha, T.: Modeling and Simulation of Bioreactor Process Dynamics. Vol. 30, p. 91

Kochba, J. see Spiegel-Roy, P. Vol. 16, p. 27

Kondratieva, E. N., Gogotov, I. N.: Production of Molecular Hydrogen in Microorganism. Vol. 28, p. 139

Koplove, H. M., Cooney, C. L.: Enzyme Production During Transient Growth. Vol. 12, p. 1

Kopperschläger, G., Böhme, H.-J., Hofmann, E.: Cibacron Blue F3G-A and Related Dyes as Ligands in Affinity Chromatography. Vol. 25, p. 101

Kosaric, N., Asher, Y.: The Utilization of Cheese Whey and its Components. Vol. 32, p. 25

Kosaric, N., Duvnjak, Z., Stewart, G. G.: Fuel Ethanol from Biomass Production, Economics, and Energy. Vol. 20, p. 119

Kosaric, N. see Magee, R. J. Vol. 32, p. 61

Kosaric, N., Wieczorek, A., Cosentino, G. P., Duvnjak, Z.: Industrial Processing and Products from the Jerusalem Artichoke. Vol. 32, p. 1

Kosaric, N., Zajic, J. E.: Microbial Oxidation of Methane and Methanol. Vol. 3, p. 89

Kosaric, N. see Zajic, K. E. Vol. 9, p. 57

Kossen, N. W. F. see Metz, B. Vol. 11, p. 103

Kristapsons, M. Z., see Viesturs. U. Vol. 21, p. 169

Kroner, K. H. see Kula, M.-R. Vol. 24, p. 73

Kula, M.-R. see Flaschel, E. Vol. 26, p. 73

Kula, M.-R., Kroner, K. H., Hustedt, H.: Purification of Enzymes by Liquid-Liquid Extraction. Vol. 24, p. 73

Kurtzman, C. P.: Biology and Physiology of the D-Xylose Degrading Yeast Pachysolen tannophilus. Vol. 27, p. 73

Lafferty, R. M. see Schlegel, H. G. Vol. 1, p. 143

Lambe, C. A. see Rosevear, A. Vol. 31, p. 37

Lang-Hinrichs, Ch. see Esser, K. Vol. 26, p. 143

Lee, K. J. see Rogers, P. L. Vol. 23, p. 37

Lee, Y.-H. see Fan, L. T. Vol. 14, p. 101

Lee, Y.-H. see Fan, L. T. Vol. 23, p. 155

Lee, Y. H., Fan, L. T., Fan, L. S.: Kinetics of Hydrolysis of Insoluble Cellulose by Cellulase, Vol. 17, p. 131

Lee, Y. H., Fan, L. T.: Properties and Mode of Action of Cellulase, Vol. 17, p. 101

Lee, Y. H., Tsao, G. T.: Dissolved Oxygen Electrodes. Vol. 13, p. 35

Lehmann, J. see Schügerl, K. Vol. 8, p. 63

Levitans, E. S. see Viesturs, U. Vol. 21, p. 169

Lim, H. C. see Agrawal, P. Vol. 30, p. 61

Lim, H. C. see Parulekar, S. J. Vol. 32, p. 207

Linko, M.: An Evaluation of Enzymatic Hydrolysis of Cellulosic Materials. Vol. 5, p. 25

Linko, M.: Biomass Conversion Program in Finland, Vol. 20, p. 163

Lücke, J. see Schügerl, K. Vol. 7, p. 1

Lücke, J. see Schügerl, K. Vol. 8, p. 63

Luong, J. H. T., Volesky, B.: Heat Evolution During the Microbial Process Estimation, Measurement, and Application. Vol. 28, p. 1

Luttmann, R., Munack, A., Thoma, M.: Mathematical Modelling, Parameter Identification and Adaptive Control of Single Cell Protein Processes in Tower Loop Bioreactors. Vol. 32, p. 95

Lynn, J. D. see Acton, R. T. Vol. 7, p. 85

Magee, R. J., Kosaric, N.: Bioconversion of Hemicellulosics. Vol. 32, p. 61

Maiorella, B., Wilke, Ch. R., Blanch, H. W.: Alcohol Production and Recovery. Vol. 20, p. 43

Málek, I.: Present State and Perspectives of Biochemical Engineering. Vol. 3, p. 279

Maleszka, R. see Schneider, H. Vol. 27, p. 57

Mandels, M.: The Culture of Plant Cells. Vol. 2, p. 201

Mandels, M. see Reese, E. T. Vol. 2, p. 181

Mangold, H. K. see Radwan, S. S. Vol. 16, p. 109

Markkanen, P. see Enari, T. M. Vol. 5, p. 1

Martin, J. F.: Control of Antibiotic Synthesis by Phosphate. Vol. 6, p. 105

Martin, P. see Zajic, J. E. Vol. 22, p. 51

McCracken, L. D., Gong, Ch.-Sh.: D-Xylose Metabolism by Mutant Strains of Candida sp. Vol. 27, p. 33

Misawa, M.: Production of Useful Plant Metabolites. Vol. 31, p. 59

Miura, Y.: Submerged Aerobic Fermentation. Vol. 4, p. 3

Miura, Y.: Mechanism of Liquid Hydrocarbon Uptake by Microorganisms and Growth Kinetics. Vol. 9, p. 31

Messing, R. A.: Carriers for Immobilized Biologically Active Systems. Vol. 10, p. 51

Metz, B., Kossen, N. W. F., van Suijidam, J. C.: The Rheology of Mould Suspensions. Vol. 11, p. 103

Moo-Young, M., Blanch, H. W.: Design of Biochemical Reactors Mass Transfer Criteria for Simple and Complex Systems. Vol. 19, p. 1

Moo-Young, M. see Scharer, J. M. Vol. 11, p. 85

Munack, A. see Luttmann, R. Vol. 32, p. 95

Nagai, S.: Mass and Energy Balances for Microbial Growth Kinetics. Vol. 11, p. 49

Nagatani, M. see Aiba, S. Vol. 1, p. 31

Nakamura, I. see Kamihara, T. Vol. 29, p. 35

Neubeck, C. E. see Faith, W. T. Vol. 1, p. 77

Neirinck, L. see Schneider, H. Vol. 27, p. 57

Nyeste, L., Pécs, M., Sevella, B., Holló, J.: Production of L-Tryptophan by Microbial Processes, Vol. 26, p. 175

Nyiri, L. K.: Application of Computers in Biochemical Engineering. Vol. 2, p. 49

O'Driscoll, K. F.: Gel Entrapped Enzymes. Vol. 4, p. 155

Oels, U. see Schügerl, K. Vol. 7, p. 1

Okabe, M. see Aiba, S. Vol. 7, p. 111

Olson, N. F. see Finocchiaro, T. Vol. 15, p. 71

Pace, G. W., Righelato, C. R.: Production of Extracellular Microbial. Vol. 15, p. 41

Parisi, F.: Energy Balances for Ethanol as a Fuel. Vol. 28, p. 41

Parulekar, S. J., Lim, H. C.: Modelling, Optimization and Control of Semi-Batch Bioreactors. Vol. 32, p. 207

Pécs, M. see Nyeste, L. Vol. 26, p. 175

Phillipson, J. D. see Anderson, L. A. Vol. 31, p. 1

Pitcher Jr., W. H.: Design and Operation of Immobilized Enzyme Reactors. Vol. 10, p. 1

Potgieter, H. J.: Biomass Conversion Program in South Africa. Vol. 20, p. 181

Poweigha, T. see Kleinstreuer, C. Vol. 30, p. 91

Quicker, G. see Schumpe, A. Vol. 24, p. 1

Radlett, P. J.: The Use Baby Hamster Kidney (BHK) Suspension Cells for the Production of Foot and Mouth Disease Vaccines. Vol. 34, p. 129

Radwan, S. S., Mangold, H. K.: Biochemistry of Lipids in Plant Cell Cultures. Vol. 16, p. 109

Ramkrishna, D.: Statistical Models of Cell Populations. Vol. 11, p. 1

Rapoport, A. J. see Beker, M. J. Vol. 35, p. 127

Reese, E. T. see Faith, W. T. Vol. 1, p. 77

Reese, E. T., Mandels, M., Weiss, A. H.: Cellulose as a Novel Energy Source. Vol. 2, p. 181

Řeháček, Z.: Ergot Alkaloids and Their Biosynthesis. Vol. 14, p. 33

Rehm, H.-J., Reiff, I.: Mechanisms and Occurrence of Microbial Oxidation of Long-Chain Alkanes. Vol. 19, p. 175

Reiff, I. see Rehm, H.-J. Vol. 19, p. 175

Reinhard, E., Alfermann, A. W.: Biotransformation by Plant Cell Cultures. Vol. 16, p. 49

Reuveny, S. see Shahar, A. Vol. 34, p. 33

Richardson, T. see Finocchiaro, T. Vol. 15, p. 71

Righelato, R. C. see Pace, G. W. Vol. 15, p. 41

Roberts, M. F. see Anderson, L. A. Vol. 31, p. 1

Roels, J. A. see Harder, A. Vol. 21, p. 55

Rogers, P. L.: Computation in Biochemical Engineering. Vol. 4, p. 125

Rogers, P. L., Lee, K. J., Skotnicki, M. L., Tribe, D. E.: Ethanol Production by Zymomonas Mobilis. Vol. 23, p. 37

Rolz, C., Humphrey, A.: Microbial Biomass from Renewables: Review of Alternatives. Vol. 21, p. 1

Rosazza, J. P. see Smith, R. V. Vol. 5, p. 69

Rosevear, A., Lambe, C. A.: Immobilized Plant Cells. Vol. 31, p. 37

Sahm, H.: Anaerobic Wastewater Treatment. Vol. 29, p. 83

Sahm, H.: Metabolism of Methanol by Yeasts. Vol. 6, p. 77

Sahm, H.: Biomass Conversion Program of West Germany. Vol. 20, p. 173

Sasse, F. see Berlin, J. Vol. 31, p. 99

Scharer, J. M., Moo-Young, M.: Methane Generation by Anaerobic Digestion of Cellulose-Containing Wastes. Vol. 11, p. 85

Schlegel, H. G., Lafferty, R. M.: The Production of Biomass from Hydrogen and Carbon Dioxide. Vol. 1, p. 143

Schmid, R. D.: Stabilized Soluble Enzymes. Vol. 12, p. 41

Schneider, H., Maleszka, R., Neirinck, L., Veliky, I. A., Chan, Y. K., Wang, P. Y.: Ethanol Production from D-Xylose and Several Other Carbohydrates by Pachysolen tannophilus. Vol. 27, p. 57

Schröder, J. see Hahlbrock, K. Vol. 18, p. 39

Schumpe, A., Quicker, G., Deckwer, W.-D.: Gas Solubilities in Microbial Culture Media. Vol. 24, p. 1

Schügerl, K.: Oxygen Transfer Into Highly Viscous Media. Vol. 19, p. 71

Schügerl, K.: Characterization and Performance of Single- and Multistage Tower Reactors with Outer Loop for Cell Mass Production. Vol. 22, p. 93

Schügerl, K., Oels, U., Lücke, J.: Bubble Column Bioreactors. Vol. 7, p. 1

Schügerl, K., Lücke, J., Lehmann, J., Wagner, F.: Application of Tower Bioreactors in Cell Mass Production. Vol. 8, p. 63

Seipenbusch, R., Blenke, H.: The Loop Reactor for Cultivating Yeast on n-Paraffin Substrate. Vol. 15, p. 1

Sevella, B. see Nyeste, L. Vol. 26, p. 175

Shahar, A., *Reuveny, S.*: Nerve and Muscle Cells on Microcarriers Culture. Vol. 34, p. 33

Shanahan, J. F. see Ho, Ch. S. Vol. 35, p. 83

Shieh, W. K., Keenan, J. D.: Fluidized Bed Biofilm Reactor for Wastewater Treatment. Vol. 33, p. 131

Shimizu, S. see Yamanè, T. Vol. 30, p. 147

Shinmyo, A. see Enatsu, T. Vol. 9, p. 111

Sittig, W., see Faust, U. Vol. 17, p. 63

Skotnicki, M. L. see Rogers, P. L. Vol. 23, p. 37

Smith, M. D. see Ho, Ch. S. Vol. 35, p. 83

Smith, R. V., Acosta Jr., D., Rosazza, J. P.: Cellular and Microbial Models in the Investigation of Mammalian Metabolism of Xenobiotics. Vol. 5, p. 69

Smith, R. V., Davis, P. J.: Induction of Xenobiotic Monooxygenases. Vol. 14, p. 61

Soda, K. see Yonaha, K. Vol. 33, p. 95

Solomon, B.: Starch Hydrolysis by Immobilized Enzymes. Industrial Application. Vol. 10, p. 131

Somers, P. J. see Barker, S. A. Vol. 10, p. 27

Sonnleitner, B.: Biotechnology of Thermophilic Bacteria: Growth, Products, and Application. Vol. 28, p. 69

Spiegel-Roy, P., Kochba, J.: Embryogenesis in Citrus Tissue Cultures. Vol. 16, p. 27

Spier, R. E.: Recent Developments in the Large Scale Cultivation of Animal Cells in Monolayers. Vol. 14, p. 119

Stewart, G. G., see Kosaric, N. Vol. 20, p. 119

Stohs, S. J.: Metabolism of Steroids in Plant Tissue Cultures. Vol. 16, p. 85

Sudo, R., *Aiba, S.*: Role and Function of Protozoa in the Biological Treatment of Polluted Waters. Vol. 29, p. 117

Suijidam, van, J. C. see Metz, N. W. Vol. 11, p. 103

Sureau, P.: Rabies Vaccine Production in Animal Cell Cultures. Vol. 34, p. 111

Szczesny, T. see Volesky, B. Vol. 27, p. 101

Taguchi, H.: The Nature of Fermentation Fluids. Vol. 1,-p. 1

Tanaka, A. see Fukui, S. Vol. 17, p. 1 and Vol. 19, p. 217

Tanaka, A. see Fukui, S. Vol. 29, p. 1

Thoma, M. see Luttmann, R. Vol. 32, p. 95

Topiwala, H. H. see Harrison, D. E. F. Vol. 3, p. 167

Torma, A. E.: The Role of Thiobacillus Ferrooxidans in Hydrometallurgical Processes. Vol. 6, p. 1

Tran Than Van, K.: Control of Morphogenesis or What Shapes a Group of Cells? Vol. 18, p. 151

Tribe, D. E. see Rogers, P. L. Vol. 23, p. 37

Tsao, G. T. see Lee, Y. H. Vol. 13, p. 35
Tsao, G. T., see Chang, M. M. Vol. 20, p. 93
Tsao, G. T. see Jang, C.-M. Vol. 25, p. 1
Tsao, G. T. see Jang, C.-M. Vol. 25, p. 19
Tsao, G. T. see Jansen, N. B. Vol. 27, p. 85
Tschopp, A. see Cogoli, A. Vol. 22, p. 1

Ursprung, H.: Biotechnology: The New Change for Industry. Vol. 30, p. 3

Vanek, Z. see Hostalek, Z. Vol. 3, p. 13
Veliky, I. A. see Schneider, H. Vol. 27, p. 57
Vieregge, J. see Hahlbrock, K. Vol. 18, p. 39
Viesturs, U. E., Kristapsons, M. Z., Levitans, E. S., Foam in Microbiological Processes. Vol. 21, p.169
Volesky, B., Szczesny, T.: Bacterial Conversion of Pentose Sugars to Acetone and Butanol. Vol. 27, p. 101
Volesky, B. see Luong, J. H. T. Vol. 28, p. 1

Wagner, F. see Schügerl, K. Vol. 8, p. 63
Wandrey, Ch., Flaschel, E.: Process Development and Economic Aspects in Enzyme Engineering Acylase L-Methionine System. Vol. 12, p. 147
Wandrey, Ch. see Flaschel, E. Vol. 26, p. 73
Wang, P. J., Hu, C. J.: Regeneration of Virus-Free Plants Through in Vitro Culture. Vol. 18, p. 61

Wang, P. Y. see Schneider, H. Vol. 27, p. 57
Wang, S. S., King, C.-K.: The Use of Coenzymes in Biochemical Reactors. Vol. 12, p. 119
Weiss, A. H. see Reese, E. T., Vol. 2, p. 181
Wieczorek, A. see Kosaric, N. Vol. 32, p. 1
Wilke, Ch. R., see Maiorella, B. Vol. 20, p. 43
Wilson, G.: Continuous Culture of Plant Cells Using the Chemostat Principle. Vol. 16, p. 1
Wingard Jr., L. B.: Enzyme Engineering Col. 2, p. 1
Wiesmann, U., Binder, H.: Biomass Separation from Liquids by Sedimentation and Centrifugation. Vol. 24, p. 119
Withers, L. A.: Low Temperature Storage of Plant Tissue Cultures. Vol. 18, p. 101
Wu, Y.-T. see Graves, D. J. Vol. 12, p. 219

Yamada, Y.: Photosynthetic Potential of Plant Cell Cultures. Vol. 31, p. 89
Yamanè, T., Shimizu, S.: Fed-batch Techniques in Microbial Processes. Vol. 30, p. 147
Yarovenko, V. L.: Theory and Practice of Continuous Cultivation of Microorganisms in Industrial Alcoholic Processes. Vol. 9, p. 1

Yonaha, K., Soda, K.: Applications of Stereoselectivity of Enzymes: Synthesis of Optically Active Amino Acids and α-Hydroxy Acids, and Stereospecific Isotope-Labeling of Amino Acids, Amines and Coenzymes. Vol. 33, p. 95

Zajic, J. E. see Kosaric, N. Vol. 3, p. 89
Zajic, J. E. see Jack, T. R. Vol. 5, p. 125
Zajic, J. E., Kosaric, N., Brosseau, J. D.: Microbial Production of Hydrogen. Vol. 9, p. 57

Zajic, J. E., Inculet, I. I., Martin, P.: Basic Concepts in Microbial Aerosols. Vol. 22, p. 51
Zlokarnik, M.: Sorption Characteristics for Gas-Liquid Contacting in Mixing Vessels. Vol. 8, p. 133
Zlokarnik, M.: Scale-Up of Surface Aerators for Waste Water Treatment. Vol. 11, p. 157